DROUGHT

DROUGHT

An Interdisciplinary Perspective

BENJAMIN I. COOK

Columbia University Press
New York

COLUMBIA
UNIVERSITY
PRESS

Columbia University Press gratefully acknowledges the generous support for this book provided by Publisher's Circle member Stephen H. Case.

Columbia University Press
Publishers Since 1893
New York   Chichester, West Sussex
cup.columbia.edu

Library of Congress Cataloging-in-Publication Data
Names: Cook, Benjamin I., author.
Title: Drought : an interdisciplinary perspective / Benjamin I. Cook.
Description: New York : Columbia University Press, [2019] |
Includes bibliographical references and index. Identifiers: LCCN
2018040215| ISBN 9780231176880 (cloth) | ISBN 9780231176897 (pbk.) |
ISBN 9780231548908 (e-book)
Subjects: LCSH: Droughts.
Classification: LCC QC929.24 .C665 2019 | DDC 577.2/2—dc23
LC record available at https://lccn.loc.gov/2018040215

Cover design: Milenda Nan Ok Lee
Cover photo: Scott London / © Alamy

Contents

Preface

rought is an *extreme* event, a relatively rare occurrence that none-theless can have widespread and devastating consequences for people and ecosystems. Compared to many climate and weather extremes, however, drought events have an impact that may not always be immediate or obvious. A wildfire, for example, can consume a housing development or an entire forest in a matter of hours or days, whereas intense precipitation and floods can destroy a bridge or road in minutes. Droughts, by contrast, are "slow-moving" disasters, with effects that accumulate incrementally over weeks, months, and even years as moisture deficits propagate through ecosystems and the hydrologic cycle. Despite this difference in temporal scale, drought impacts on agricultural productivity, ecosystem health, and even human morbidity make these events among the most expensive and damaging climate extremes in the world.

Indeed, one of the worst natural disasters in the history of the United States was the Dust Bowl drought of the 1930s. This nearly decade-long event devastated the center of the country, created almost unprecedented levels of land degradation and wind erosion, and caused the mass migration of millions of people and the abandonment of thousands of farms (B. Cook et al., 2009; Hansen & Libecap, 2004; Lee & Gill, 2015; Schubert et al., 2004). Further back, history is replete with examples of droughts contributing to the collapse of ancient societies, including the Ancestral Puebloans of southwestern North America (Benson, Petersen, & Stein, 2007), the Maya in Central America (Medina-Elizalde & Rohling, 2012), and the Angkor in modern-day Cambodia (Buckley et al., 2010). Even modern agricultural and political systems, with all our technological advances, can show remarkable vulnerability to drought. The recent drought in the eastern Mediterranean has been linked to the political instability in the region and has even been raised as a possible contributing factor to the Syrian civil war (Gleick, 2014; Kelley et al., 2015). A drought that only recently ended in California (as of this writing in spring 2018) had devastating consequences for the state's agriculture, ecosystems, and water resources (He et al., 2017; Howitt et al., 2015). In some cases, drought impacts may even be realized globally. In the early 2000s, for example, a series of contemporaneous regional droughts suppressed vegetation productivity at a magnitude sufficient to cause a global-scale reduction in the terrestrial carbon sink (Zhao & Running, 2010).

Comprehensive discussions of drought dynamics and impacts are often difficult, however, even in the academic literature and especially within the context of climate change. Much of this difficulty arises because of the fundamentally interdisciplinary nature of drought, a phenomenon sitting at the intersection of climatology, meteorology, hydrology, ecology, agronomy, and even economics. Different disciplines will frequently view the same drought from different perspectives, often with little effort made to synthesize understanding of a given event across the entire system. A meteorologist, for example, might investigate the precipitation deficits and associated atmospheric circulation anomalies that caused a drought. Alternatively, a hydrologist might choose to focus on soil moisture or streamflow during the same drought, variables where additional processes (e.g., infiltration, runoff, and evapotranspiration) are important. Neither approach is wrong, per se, but such disciplinary separation can result in biased interpretations and sometimes opposing conclusions about the same event simply because the definition of *drought* is different in different fields.

As a result, there are no straightforward answers to some of the most pressing questions surrounding drought and climate change: How does this drought compare to droughts in the past? Is climate change making droughts worse? How do we know when a drought is over? Will there be enough water for everyone in the future? My goal with this book is to provide an accessible resource to students and researchers across environmental disciplines, show-casing drought as a unique interdisciplinary phenomenon. Because of my own background, much of the focus will be on drought and hydroclimate from a climatological perspective. Included are discussions of how the global climate system shapes patterns of aridity and regional drought variability, a survey of major drought events over the past 12,000 years, and an explora-tion of the effects of climate change on drought dynamics now and in the future. Beyond this, however, I will explore the impacts of drought on water resources, people, and ecological systems. Thus, I include cases studies of important past events (the Dust Bowl drought of the 1930s and the Sahel drought of the 1970s), an examination of the role of drought in desertifica-tion and land degradation, and a review of the development of irrigation and groundwater exploitation as tools to address water shortages.

Given the breadth of the topic and the targeted audience for this book, I will by necessity simplify certain discussions and presentations (e.g., the description of the Hadley circulation) and instead focus on the points most relevant for drought and hydroclimate. Drought, ultimately, is a complex concept that can be defined in a variety of ways. No two droughts are the same across regions or time, and the severity of their impacts, now and in the future, will depend on many different physical, biological, and social factors. This book will capture such diversity and showcase the importance of embracing this broad perspective as we consider how to make sense of droughts in a world that, because of climate change and development, will look very different in the future compared to the past.

Acknowledgments

This book is easily the most difficult writing task I have ever undertaken. (Even my dissertation felt easier.) Few things are quite as intimidating as staring at a blank page and knowing it is up to you to fill those pages with useful insights, edit them down to something approaching sensibility, and decide when it is finally done. It is an immense privilege. Few people have the opportunity and platform to synthesize and share their thoughts and ideas surrounding the topics they have dedicated much of their career to understanding. It is beyond humbling. Also, it's nice having a place to write down all the water cycle numbers that I was tired of looking up in the literature.

Thank you to Patrick Fitzgerald for first approaching me with the idea to write a book on drought, and for remaining endlessly patient as I slogged away at it these last few years. Thank you to Kevin Anchukaitis, Ron Miller, Michael Puma, Richard Seager, Jason Smerdon, Park Williams, Lizzie Wolkovich, and all my other collaborators past and present who have made me a better scientist. Thank you as well to Michael Mann, Howie Epstein, and Paolo D'Odorico for taking a chance on me at the University of Virginia when nobody else would.

And thank you to my parents, Rori, and Luca for their endless love and patience.

DROUGHT

ONE

Introduction to the Hydrologic Cycle and Drought

W ater is an integral component of nearly every human and natural system on Earth, a planet unique in our solar system for its abundance of water at the surface in all three phases (liquid water, gaseous water vapor, and solid ice). Water facilitates the movement of energy and materials, provides the basis for all biological life, and serves as a critical resource for most human activities. Water is not, however, evenly distributed, and various processes affect the availability of water for people and ecosystems across seasons, regions, and time periods. Here, we review some of the most important and fundamental concepts related to the hydrologic cycle and distribution of water on Earth. How much water is there, and where is it located? What processes are responsible for moving this water around the Earth system? How much of this water is available for, and used by, people? And, finally, how do we define and measure drought?

The Global Hydrologic Cycle

The global hydrologic (water) cycle (figure 1.1) describes the continuous movement of water between and through **reservoirs** in the Earth system. A reservoir is any physical subdivision that contains an amount of water we can measure, whether globally (e.g., the global atmosphere or ocean) or locally (e.g., any individual aquifer or lake). Water moves between these reservoirs as

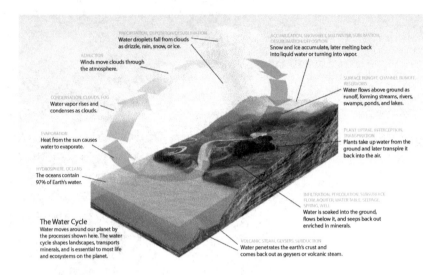

FIGURE 1.1 The global hydrologic (water) cycle. Water is stored in the atmosphere as water vapor (the gaseous phase) and in clouds as water (the liquid phase) and ice (the solid phase). The atmosphere transports this water from region to region, gaining water from the surface through evapotranspiration of liquid water and sublimation of snow and losing it through precipitation processes (e.g., rain and snow). Water at the surface is stored as liquid water in surface reservoirs (oceans, lakes, rivers, etc.), soils, groundwater, and vegetation and as solid snow and ice in glaciers, ice sheets, sea ice, and permafrost soils. At the surface, ice and snow can melt, and the resulting liquid water can recharge moisture in dry soils before it eventually runs off into rivers, lakes, and oceans. Climate, ecology, human activities, and landscape processes all play critical roles in nearly all aspects of the water cycle, including how much water is available for use by people and ecosystems. *Source:* Ehud Tal, CC BY-SA 4.0.

fluxes that may involve transformations of water from one phase to another. For example, liquid water evaporates from the surface, becoming water vapor in the atmosphere. This water vapor, in turn, eventually condenses to ice and liquid water, forming clouds and finally returning to the surface as precipitation (e.g., rain and snow). All phase transformations of water involve exchanges of energy with the environment, making water critically important for modulating the movement of energy through the Earth system. Estimates of the global volumes of water in the major Earth system reservoirs are shown in table 1.1, and the fluxes between major reservoirs are shown in table 1.2. With regard to water, the Earth can be considered a "closed" system, with no significant additions or losses of water over time.

TABLE 1.1
Volumes of Water Present in the Major Earth System Reservoirs

Reservoir	Volume (km³)	Residence time	Notes
Total	1,386,000,000 to 1,460,000,000		Earth system, fresh and saline, all phases
Oceans	1,338,000,000 to 1,400,000,000	3,000 to 3,230 years	~97% of all water on Earth
Atmosphere (liquid water and ice in clouds and water vapor)	12,000 to 15,000	9 to 10 days	
Cryosphere (ice and snow, perennial and seasonal)	43,400,000		
Perennial ice and snow (ice caps and glaciers)	24,064,000 to 29,000,000		~68.7% of freshwater, 2% of total Earth water
Greenland and Antarctic ice	32,000,000		
Mountain glaciers	100,000	20 to 100 years	
Groundwater (saline and fresh)	23,400,000	100 to 200 years (shallow); 10,000 years (deep)	
Groundwater (fresh)	10,530,000		~30% of global freshwater
Freshwater (all phases)	35,030,000		~2.5% to 3% of all water on Earth

Note: These values reflect global average conditions across time and space. Of all the water on Earth, only about 2.5% is fresh, and less than 1% is fresh and available at the surface in liquid form. Ranges reflect uncertainties or variability in the observational estimates for various reservoirs.

Source: UCAR Center for Science Education (UCAR, 2011).

If we know the size of a reservoir (i.e., the total amount of water it contains) and the fluxes into and out of that reservoir, we can calculate the **residence time**. Residence time represents the average amount of time a particle (in our case, any given water molecule) spends in a particular reservoir. This tells us how long it typically takes for the reservoir to replace its entire mass. For example, the atmosphere contains approximately 13,500 km³ of water

TABLE 1.2
Fluxes Between Reservoirs in the Earth System

Flux	Volume (km³/year)
Precipitation (total)	505,000
Ocean precipitation	398,000
Land precipitation	96,000 to 107,000
Evapotranspiration (total)	505,000
Ocean evaporation	434,000
Land evaporation (from bare soil and leaf surfaces)	50,000
Transpiration (uptake by plants)	21,000
Runoff (from land to ocean)	36,000
Melting (transitioning ice/snow to liquid water on the land surface)	11,000

Note: These values reflect long-term global averages.
Source: UCAR Center for Science Education (UCAR, 2011).

(the average of the range in table 1.1) in the form of clouds (liquid water and ice) and gaseous water vapor. The atmosphere gains water as evapotranspiration from the land surface and oceans at a rate of 505,000 km³/year (table 1.2). We can calculate the residence time for water in the atmosphere as

$$\frac{water\ in\ the\ atmosphere\ (km^3)}{global\ evaporation\ rate\ \left(^{km^3}\!/_{yr}\right)}$$

$$\frac{13,500\ (km^3)}{505,000\left(^{km^3}\!/_{yr}\right)}$$

$$\sim 0.0267\ yrs = \sim 9.75\ days$$

Replacing the total volume of water in the atmosphere takes 9 to 10 days on average, a fairly short turnover time (in reality, the residence time is even shorter, ~3 days, because air-surface exchange processes occur more rapidly than the global evapotranspiration rate would suggest).

Other reservoirs have much longer residence times. The global ocean contains ~1,339,000,000 km³ of water (the average of the range in table 1.1), and the average annual evaporation rate from the ocean is ~434,000 km³/year (table 1.2). We can calculate the residence time for the global ocean as

$$\frac{water\ in\ the\ ocean\ (km^3)}{ocean\ evaporation\ rate\ \left(\frac{km^3}{yr}\right)}$$

$$\frac{1,339,000,000\ (km^3)}{434,000\ \left(\frac{km^3}{yr}\right)}$$

$$\underline{\sim 3,085\ years}$$

Replacing the total volume of water in the global ocean thus takes over 3,000 years, a much longer time than for total replacement of water in the atmosphere.

Integrated across all phases and the entire Earth system, the total amount of water on Earth is estimated at 1,386,000,000 to 1,460,000,000 km³, almost all of which (~97 percent) resides in the oceans. Total freshwater (in all phases, including the cryosphere) accounts for only a tiny fraction of the total water on Earth (~2.5 to 3 percent). Most of the global freshwater resources (~69 percent of total freshwater) are locked in perennial ice and snow, such as mountain glaciers and the ice sheets on Antarctica and Greenland. Ultimately, less than 1 percent of water on Earth is fresh *and* present in liquid form in natural and artificial surface reservoirs (e.g., lakes, swamps, and rivers), with ~87 percent of surface freshwater stored in lakes. Below the surface, **groundwater** is stored in underground **aquifers** composed of permeable rock or unconsolidated geologic material (e.g., gravel and sand), providing another important source of freshwater (~30 percent of global freshwater is stored as groundwater). Groundwater dynamics vary widely from region to region, however, and are strongly controlled by local geology, ecology, and climate. Depending on these conditions, aquifers may be recharged from the surface fairly regularly (with residence times of decades to centuries), or they may be largely disconnected from the hydrologic cycle with very small rates of recharge or discharge (with residence times of millennia or longer). For the latter, these water resources are typically referred to as *fossil groundwater* because they are effectively nonrenewable on the typical timescales of human exploitation.

Moisture Processes in the Atmosphere and at the Surface

The horizontal and vertical movement of water within and between regions and reservoirs is strongly controlled by different processes in the climate system. Because continental to global water budgets must satisfy mass balance (inputs must balance outputs), they are relatively well constrained and are generally in equilibrium over timescales of a year or more. Shifts in the climate system, however, can cause significant short-term deviations in the local water balance. These deviations include seasonal differences in moisture availability, such as dry and wet seasons in regions with monsoon and Mediterranean climates, and droughts and pluvials that may persist for a year or more. Understanding any of these deviations thus requires understanding the fundamental physical processes that describe the moisture balance in the climate system. A simplified version of the moisture balance equation for the atmosphere (valid for time averages) is

$$\overline{ET} - \overline{P} = \nabla \cdot \overline{Q} + \overline{\tfrac{\partial w}{\partial t}}$$

where $\overline{ET}$ is evapotranspiration from the surface, $\overline{P}$ is precipitation, $\nabla \cdot \overline{Q}$ is horizontal moisture divergence, and $\overline{\tfrac{\partial w}{\partial t}}$ is the change in moisture content in the air over time. For the land surface, the moisture balance equation is

$$\tfrac{\partial s}{\partial t} = P - ET - R$$

where $\tfrac{\partial s}{\partial t}$ is the rate of change in moisture storage in the soil, P is precipitation, ET is evapotranspiration, and R is **runoff** (including at the surface and belowground).

The *moisture divergence* term describes the net horizontal movement of water in the atmosphere. Positive divergence indicates the net export of water out of a region or location, whereas negative divergence (*convergence*) indicates the net accumulation of moisture. Moisture divergence and convergence are primarily controlled by horizontal and vertical circulation processes in the atmosphere, operating on local to global scales. In the atmosphere, the net exchange of water between the atmosphere and the surface $(\overline{ET} - \overline{P})$ must be balanced by changes in the moisture content in the atmosphere $(\overline{\tfrac{\partial w}{\partial t}})$

and the inflow or outflow of water vapor $(\nabla \cdot \overline{Q})$. At the surface, changes in the rate of moisture storage in the soil $(\frac{\partial s}{\partial t})$ are governed by inputs through **precipitation** (P) and exports from evapotranspiration (ET) and runoff (R). **Evapotranspiration** includes the physical processes of *evaporation* from the ocean and land surface (over land, this occurs from water on the surface of soils or plants) and *transpiration*, the flux associated with water taken up by plant roots and lost from the plant to the atmosphere. This gas exchange occurs primarily through leaves via stomata, openings that permit carbon dioxide to diffuse into the leaf for photosynthesis and simultaneously allow water vapor to diffuse out of the plant into the atmosphere. Previous, more conservative estimates have attributed about one-third of total terrestrial evapotranspiration to transpiration (table 1.2). However, more recent analyses suggest this transpiration fraction may be much larger and likely varies significantly across ecosystems (Jasechko et al., 2013; Schlesinger & Jasechko, 2014). One such study, synthesizing evidence from across a range of modeling and observational approaches, estimated that transpiration accounts for ~57 percent of global terrestrial evapotranspiration, with significant ranges across vegetation types: ~31 percent for wetlands and over 60 percent for most forests, shrublands and grasslands, and crops (Wei et al., 2017). Transpiration is also estimated to return ~39 percent of terrestrial precipitation to the atmosphere, an important flux in the global water cycle (Schlesinger & Jasechko, 2014).

Transpiration dominates over physical evaporation in many vegetated regions for two reasons. First, plants have roots that can access moisture at depths, acting as conduits connecting these deeper pools to the atmosphere and increasing the total amount of water available for evapotranspiration. Roots in tropical rainforests, for example, can extend ~20 m or more, and roots in many grasslands and temperate forests can reach depths of ~5 m (Canadell et al., 1996). Second, because transpiration occurs from the surfaces of leaves and many plants have overlapping layers of leaves in the canopy, plants increase the total effective area from which evapotranspiration can occur. For example, forests have a typical leaf area index (LAI) of ~5 m^2 m^{-2} (translating to 5 m^2 of leaves for every 1 m^2 of ground), whereas grasslands and shrublands typically have lower LAI values of 1 to 3 m^2 m^{-2} (Asner et al., 2003). In both cases, this translates to an increase in evaporative surface area compared to the bare ground alone.

The total amount of evapotranspiration that occurs is controlled by the availability of moisture at the surface and the evaporative demand from the atmosphere. Evaporative demand is often referred to as *reference* or **potential evapotranspiration (PET)**, defined as the total amount of evapotranspiration that would occur if the moisture supply at the surface was not a limiting factor. When combined with precipitation, PET is a useful variable for characterizing the relative *aridity* (or dryness) of different regions (see chapter 2). PET increases with temperature, wind speed, and available energy (e.g., insolation from the sun) and decreases with increasing humidity (water vapor content of the atmosphere). Temperature is often the dominant control because as temperature rises, it causes an exponential increase in the *saturation vapor pressure* (the point at which water vapor is in equilibrium with a liquid water surface; effectively the maximum water vapor–holding capacity of the atmosphere). This follows the well-known **Clausius-Clapeyron relationship**: for every 1°C or 1K of warming, the saturation vapor pressure increases by ~7 percent.

In regions where surface moisture is not limiting (primarily the oceans), evapotranspiration will steadily increase in response to warmer temperatures and the associated increase in evaporative demand. Over land areas, however, only a finite amount of water is available at the surface, and, as this reservoir is drawn down, evapotranspiration will similarly decline, even if demand continues to increase. In densely vegetated regions, this occurs primarily through downregulation by the vegetation. As soils begin to dry, plants will progressively close their stomata to limit water losses and thus reduce the risk of cavitation (a catastrophic loss of water tension in plant vessels, often causing significant structural damage). This closure of the stomata also limits the diffusion of carbon dioxide into the leaves for photosynthesis, decreasing plant productivity. Managing this trade-off between water loss and carbon gain is one of the major challenges facing plants as they deal with water deficits during drought events.

Precipitation refers to all particles of water in its liquid phase (e.g., rain) and solid phase (e.g., snow and ice) that originate in the atmosphere and fall to the Earth's surface. For most terrestrial regions, this is the ultimate source of moisture supply to soils, ecosystems, lakes and rivers, and other surface reservoirs. Not all precipitation that falls, however, is available to recharge these terrestrial reservoirs. In densely vegetated areas (e.g., forests), a large fraction of precipitation is *intercepted* by plant canopies, accumulating on leaves and stems before being evaporated directly back to the atmosphere. Evaporation of intercepted water differs critically from transpiration because the intercepted water is unavailable for plant use and is thus effectively lost back to the atmosphere. The remaining fraction either passes through the canopy unhindered (*throughfall*) or flows along and drips off plant surfaces to the ground (*stemflow*) where it can collect on the land surface, flow across the surface, or infiltrate into the soil. In some ecosystems, fog and cloud water collection by plants, referred to as occult precipitation, can be a significant source of moisture input in addition to precipitation (Weathers, 1999). Plants likely benefit from occult precipitation in a few ways: by taking it up directly, using it to reduce transpiration losses during photosynthesis, or collecting it on plant surfaces and channeling it to the soils (Weathers, 1999). Occult precipitation has been demonstrated to be especially important in exceptionally foggy ecosystems and arid regions with small precipitation inputs, including the coastal redwood forests of Northern California (Dawson, 1998), the high-altitude Páramos grassland ecosystems in the Tropical Andes (Cárdenas et al., 2017), and the Cape Floristic Region of South Africa (Nyaga et al., 2015).

Liquid water penetrates the soil profile through the process of **infiltration**, which is largely a function of the physical characteristics of the soil. Water infiltrates most easily into less dense soils with high porosity (total empty space), determined by the geometry, size, and arrangement of soil particles. Porosity is highest in soils with coarse texture (sand) and fine texture (clay) and lowest at midtextures (silt). Adding organic matter to soils lowers the bulk soil density, creates empty space, and increases porosity, as do biological activities (e.g., root growth and animal burrowing). Some soils also have layers of impermeable rock that may limit infiltration at the surface and deeper in the soil profile. The infiltration can similarly be stymied if intense precipitation quickly overcomes the infiltration capacity of the soil, if the freezing of water in the soil creates a temporary impermeable barrier, or if the soil is saturated and cannot accommodate any more water. In such cases, water will not infiltrate and will instead flow horizontally as *runoff*.

Runoff is a critical component of the surface water budget, representing the excess of water that is available after accounting for the moisture supply (precipitation) and demand (evapotranspiration) components of the surface moisture budget. Surface and subsurface runoff is the main avenue by which many surface reservoirs (lakes and rivers) and groundwater reservoirs are recharged. The main measure of the contribution of precipitation to runoff is *runoff efficiency*, effectively the amount of runoff generated per unit of precipitation (McCabe & Wolock, 2016). Runoff efficiency is affected by a variety of processes that can change over time (precipitation intensity, evapotranspiration, and land cover) and reflects the amount of precipitation that can recharge water resources at the surface. Continental wetness is defined using a similar metric, the *runoff ratio*, which is the ratio of runoff to precipitation integrated over the entire continent. Africa, home of the largest desert region in the world, is the driest continent by this metric, with a runoff ratio of ~16 percent, whereas most other continents have runoff ratios of ~30 to 50 percent.

In colder areas (at high latitudes and in mountainous regions), the seasonal cycle and variability in runoff and streamflow are closely connected to snowpack dynamics. In these regions, winters are usually continuously cold enough that most precipitation falls as snow, which accumulates at the surface over the course of the winter and typically melts rapidly over several days to weeks in a single large pulse in the spring. This rapid melt recharges soil moisture and often overcomes the infiltration capacity of the soils, generating runoff and streamflow. This dynamic plays out in many of the most important river basins in the world, including the Colorado River in western North America, the Upper Mekong River in Southeast Asia, the Columbia River in the Pacific Northwest, and the upper basin of the Rio Grande.

Globally, $\overline{ET} - \overline{P}$ varies greatly across latitudes and regions and is a useful indicator of the mean hydroclimate state at a given location or time of year. Because changes in $\overline{ET} - \overline{P}$ must be balanced by changes in moisture divergence, strongly negative values indicate a net convergence of moisture in wet regions like the tropics and midlatitudes and positive values a net divergence in drier areas, such as the subtropical latitudes. Broadly, these geographic differences are a consequence of the general circulation of the atmosphere (discussed in chapter 2). Globally, most evapotranspiration occurs over the oceans, and, averaged over a year, the atmosphere is a net exporter of moisture from the oceans to the continents. Evaporation from the oceans is the source of approximately one-third of precipitation over land, which then either returns to the atmosphere via evapotranspiration from the land or returns to the oceans via horizontal runoff flowing into rivers, over land, or underground.

Human Water Use

Large-scale human manipulation of the hydrologic cycle and appropriation of water resources began with the agricultural revolution ~10,000 years ago. Reliable sources of water were required to support agriculture and the increasingly sedentary human populations that established permanent settlements. These manipulations varied in both size and complexity. Some of the oldest wells in the world have been dated to ~6500 BCE in Israel. In 312 BCE, ancient Romans constructed the Aqua Appia, the first aqueduct to supply and transport water to Rome from a source 16.4 km away. The Angkor civilization of Southeast Asia used massive *barays* (surface reservoirs) to store water for irrigation and flood management. Technological innovations since have allowed for larger-scale manipulations of the hydrologic cycle and appropriation of water from sources that were previously inaccessible (e.g., deep groundwater aquifers). The fundamental goal, however, remains the same: to access and store freshwater for human use.

Exploitation of water resources can be divided into two categories: *withdrawal* and *consumption*. Withdrawal refers to the total amount of water taken from a source, such as a lake or groundwater aquifer. A fraction of this water may return to the source (e.g., as runoff from an agricultural field) to be used again. Consumption is the amount of water that is used but no longer available. This includes water lost to the atmosphere through evapotranspiration by vegetation and water incorporated into the physical structure (biomass) of crops. It is estimated that 12,500 to 14,000 km^3 of freshwater are available globally for human consumption each year (Jackson et al., 2001). Over the last century, global water withdrawals in all sectors (domestic/municipal use, agriculture, industry, and evaporation from artificial reservoirs) have increased eightfold, from ~500 km^3 yr^{-1} in 1900 to over

4,000 km³ yr⁻¹ in 2010 (FAO, 2016; Shiklomanov, 2000), with consumption accounting for ~50 to 60 percent of global water withdrawals.

Modern appropriations of water by people can be divided into two categories: **blue water** and **green water**. The traditional sphere of human water resource management has focused on blue water (39 percent of global human water use), referring to resources contained in reservoirs such as rivers, lakes, wetlands, groundwater, and artificial impoundments (e.g., lakes created by dams). Blue water originates mostly from runoff, and ~54 percent of it is globally accessible to human use (Postel et al., 1996). These resources have traditionally been the easiest to control and manage proactively through various technologies and tools, including groundwater pumping, dams and reservoirs, and other infrastructure improvements. Green water (61 percent of global human water use) is contained within the soils, recharged by precipitation, and used by natural vegetation and crops before being evapotranspired back to the atmosphere. Green water is more difficult to manage proactively and is more sensitive to climate and weather variability and extremes. Regardless, green water availability is critically important for the ~80 percent of global agricultural areas that are not irrigated and thus solely dependent on rainfall.

Human water withdrawals fall into three broad categories: agricultural (~69 percent), industrial (19 percent), and municipal (12 percent) (FAO, 2016). The agricultural component includes water used for livestock and aquaculture, but by far the largest agricultural use is the irrigation of crops for human consumption and animal fodder. Only ~20 percent of global croplands are irrigated, but these areas account for ~40 percent of global crop production. The proportion of water withdrawn for agriculture varies by continent and makes up a greater proportion over Asia and Africa, where agriculture is a greater fraction of the total economy. In terms of absolute water amounts, most irrigation occurs in Asia, which hosts 78 percent of the world's irrigated croplands. Industrial water use makes up the next largest fraction of human withdrawals, with most of this water stored in aboveground reservoirs and used for hydropower and storage. The remaining fraction is municipal or domestic water use, including direct water consumption by people: water used for showering and bathing, cooking, and other domestic activities.

What Is a Drought?

Fundamentally, a **drought** is a period of moisture deficit relative to some baseline average or normal state (the opposite of a drought, a period of moisture surplus, is a *pluvial*). This normal state is always defined based on the local climate, so that a drought in a relatively arid region (e.g., Arizona) does not translate to the same absolute water deficit as a drought in a wetter region (e.g., Massachusetts). Droughts represent a temporary deviation from this normal baseline, contrasting with **aridity**, a permanent state of the climate typically associated with regions of low absolute precipitation (Wilhite, 2000).

Within this broad definition, however, droughts encompass a diversity of phenomena affecting all aspects of the hydrologic cycle, with dynamics and impacts that vary substantially over time and space (Wilhite, 2000; Wilhite & Glantz, 1985). Droughts can be measured in terms of precipitation, soil moisture, or streamflow. They may be short-lived events that last for weeks or months or persistent extremes that continue for years or even decades. They may be exceptionally rare or occur with alarming regularity. Their impacts on ecosystems, agriculture, and societies may be modest, lasting only a season or two, or catastrophic, causing a wholesale reorganization of ecosystems and societies. Droughts may be extremely localized, covering tens to hundreds of square kilometers and impacting only local communities, or they may extend across entire continents. This diversity of drought dynamics and impacts ultimately precludes the development of a single unifying theory of drought. More importantly, the inherently interdisciplinary nature of drought can present significant challenges for predicting drought events and mitigating their effects on people and ecosystems.

Drought Monitoring

Monitoring and studying droughts requires quantitative indicators to track the onset, severity, evolution, and eventual termination of drought events. Because of the complexity of how droughts are expressed spatially, temporally, and across the hydrologic cycle, no single drought indicator is appropriate for all situations. Further, certain types or characteristics of drought may be more difficult to track than others because of data limitations. In many regions, for example, high-quality records of precipitation and temperature are available for a century or more; analyzing droughts in terms of these variables is therefore reasonably straightforward. Information on other drought-relevant variables (e.g., snow cover, humidity, soil moisture, runoff, and evapotranspiration), however, is often limited or even nonexistent for many regions. In such cases, researchers must derive or estimate these quantities using other datasets that are more readily available. For example, soil moisture can be estimated using a simple "bucket" model, where variability in soil moisture is calculated from the balance of precipitation inputs and evapotranspiration losses (the latter often calculated as a function of temperature). Additionally, more sophisticated hydrologic models can be forced offline (i.e., applied using only inputs of atmospheric variables) to generate an observation-constrained suite of estimates for different hydrological variables (e.g., runoff, soil moisture, and evapotranspiration).

Drought indicators typically fall into two categories. The first is standardized indices, which express drought severity and hydroclimate variability in anomalous terms relative to the defined normal baseline. Classic examples of these include the Standardized Precipitation Index (SPI), which tracks precipitation variability; the Palmer Drought Severity Index (PDSI), which estimates soil moisture variability; or the Standardized Runoff Index (SRI), which indicates variability in runoff and streamflow. Droughts can also be analyzed from the perspective of the base datasets themselves (precipitation and snow cover), with the results expressed as anomalies or percentages above or below the long-term normal baseline. Similar approaches can be used to track drought impacts: satellite estimates of anomalies in vegetation growth and productivity are often used to quantify ecosystem responses to drought. Standardized indices have the advantage of allowing for easy comparisons of droughts across regions or individual events and are widely employed in the fields of climatology, hydrology, and ecology.

The practical application of standardized drought indices for resource management, however, is limited. From a management perspective, absolute volumes of water are critical for determining if water deliveries can meet demands. To address this need, the second category of drought indicators

uses predefined thresholds, below which water deficits begin to have significant consequences (Van Loon, 2015). These threshold levels are location and
system specific and reflect the volumes of water needed for activities such
as irrigation, river management for navigation and wildlife, and municipal
needs. Lake Mead, for example, is the largest surface reservoir in the western
United States, providing water from the Colorado River to over 20 million
people. This reservoir has a peak elevation (maximum capacity) of ~1,221 feet
and a "dead pool" level of 895 feet, the elevation at which no water will flow
downstream out of the lowest water outlet of the Hoover Dam (National
Park Service, 2015). Water levels falling below this dead pool would have significant repercussions for hydropower generation from the Hoover Dam and
water resource availability to Colorado River stakeholders.

One of the more comprehensive tools developed for tracking and monitoring drought conditions is the U.S. Drought Monitor, or USDM (http://
droughtmonitor.unl.edu). The USDM provides, on a weekly basis, a snapshot
of current drought conditions for the United States, based on five categories
of drought from "abnormally dry" (D0) to "exceptional drought" (D4). The
USDM blends a variety of indicators measuring drought conditions across
the hydrologic cycle, including SPI, PDSI, and streamflow monitored by the
U.S. Geological Survey. It also incorporates local observations of snow water,
groundwater, reservoir levels, and vegetation conditions. The USDM thus
provides a comprehensive overview of drought conditions and impacts that
can be used to effectively monitor and track the evolution of drought from
week to week across the country.

Two examples highlight the utility of the USDM for monitoring long-term
(multiyear) and short-term (less than 1 year) droughts. The USDM shows
that California has experienced three waves of multiyear drought since the
monitor began operation in 2000, culminating in the most severe event from
2012 through 2016 (figure 1.2). The cumulative effect of these three droughts
can be seen in a variety of observations across the state, including significant
levels of groundwater depletion (Thomas et al., 2017; S. Y. S. Wang et al., 2017).
The USDM also documented the rapid development of drought conditions
in the central United States from spring into summer of 2012 (figure 1.3). This
was one of the most widespread and severe short-term droughts in the country since the 1930s. Although it lasted less than a year, the rapid development

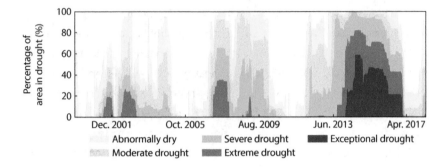

FIGURE 1.2 Cumulative area (percentage) of various categories of drought in California (2000–2018). Over this period, California experienced three widespread, multiyear drought events of varying severity: (1) 2001–2004, (2) 2007–2010, and (3) 2012–2016. *Source*: Data from the U.S. Drought Monitor, National Drought Mitigation Center at the University of Nebraska–Lincoln, https://drought.unl.edu.

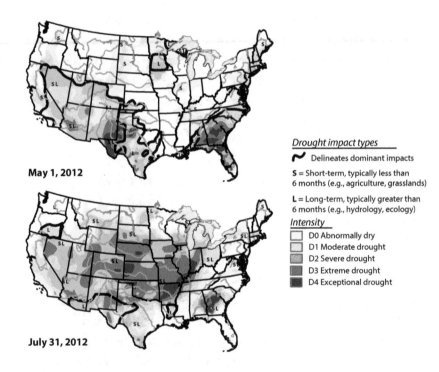

FIGURE 1.3 Drought conditions in the contiguous United States from the United States Drought monitor for May 1, 2012, and July 31, 2012. This summer saw the rapid development and intensification of severe-to-exceptional drought conditions over much of the central United States, a "flash drought" that was not predicted ahead of time. *Source*: U.S. Drought Monitor, National Drought Mitigation Center at the University of Nebraska–Lincoln, https://drought.unl.edu.

and severity of this "flash drought" meant there was little advance warning, resulting in major impacts to agriculture across the central United States (Hoerling et al., 2014).

Categories of Drought

Traditionally, droughts are categorized based on where in the hydrologic cycle the associated moisture deficits occur (Wilhite, 2000; Wilhite & Glantz, 1985), with an additional category defined based on the socioeconomic impacts of said droughts (figure 1.4). Recent years, however, have seen a shift away from the idea of *socioeconomic* as a category of drought. Instead, this new perspective recognizes that the socioeconomic consequences of drought are a function of both physical moisture deficits and human activities that may amplify or ameliorate drought impacts (AghaKouchak, 2015; Van Loon et al., 2016). Importantly, these drought categories are not independent of each other.

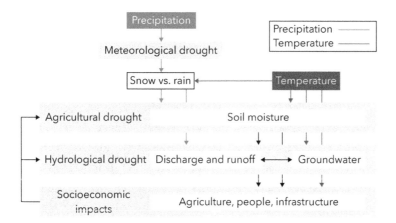

FIGURE 1.4 Categories and propagation of drought across the hydrologic cycle. Precipitation deficits (meteorological drought) are the ultimate driver of most drought events, with the associated moisture anomalies propagating into the soils (agricultural drought) and eventually into runoff, streamflow, and groundwater (hydrological drought). Temperatures can exacerbate agricultural and hydrological drought by increasing snowmelt, the fraction of precipitation falling as rain versus snow, and evaporative losses from the surface. Surface water flows and interactions between human factors and the physical environment are indicated by black arrows, highlighting the importance of two-way interactions and the capacity for human activities to exacerbate or ameliorate moisture deficits associated with physical droughts. *Source:* Adapted and modified from Figure 2, Van Loon, 2015.

Moisture deficits propagate over time across the hydrologic cycle, and important feedbacks within the Earth system and from human activities operate across a range of spatiotemporal scales. These categories should therefore be viewed as analytical conveniences used to investigate and describe drought characteristics (e.g., severity and duration), driving mechanisms (e.g., lack of precipitation and increased evapotranspiration), and impacts (e.g., soil moisture availability for crops and reservoir storage). To demonstrate these categories of drought and how they are connected, we will use California (and its most recent drought, that of 2012–2016) as a case study.

Most droughts, regardless of their eventual categorization, begin as a *meteorological drought*, a period of below-average precipitation (rain or snow). Any given day may have precipitation or not, but for meteorological droughts, the focus is on precipitation deficits that typically accumulate over the course of weeks and longer. These droughts may manifest as a reduced number of precipitation events (i.e., the frequency), reduced storm depth (i.e., the amount of precipitation generated by any given storm), or both. Because precipitation is intimately connected to processes in the climate system, regional meteorological droughts can often be attributed to shifts in various modes of climate variability, such as the El Niño Southern Oscillation or the North Atlantic Oscillation. The simplest way to investigate meteorological drought variability is to analyze precipitation data, using the *water year* (for California, October of the previous calendar year through September of the current year). The focus on water-year totals reflects the fact that most precipitation in California occurs during the cold season (October–March) and precipitation from these months is the primary source of water for human and ecosystem demands during the summer dry season. These data clearly show the 2012–2016 drought, with precipitation falling below the long-term average (grey line) in all 4 water years (figure 1.5). Looking back across the entire record, however, single and multiyear periods of lower precipitation are clearly present in the past. In part, this supports the consensus view that the precipitation deficits associated with this most recent period of California drought were likely not influenced by climate change (Seager et al., 2015).

Agricultural drought occurs when significant moisture deficits develop in the unsaturated zone of the soil above the groundwater table. These droughts

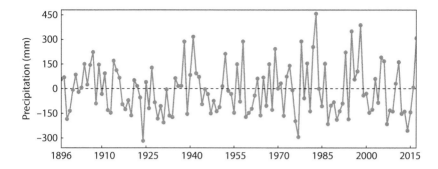

FIGURE 1.5 Precipitation anomalies (relative to a 1901–2000 baseline) for each water year (from October of the previous calendar year through September of the current year) in California from 1896 to 2017. The most recent drought event is apparent as a clear period of persistent below-average precipitation. A longer historical perspective, however, shows periods of drought in the past with durations and accumulated precipitation deficits similar to, or even exceeding, those of this most recent drought. *Source*: Data from the NOAA National Centers for Environmental Information. Published March 2018. Climate at a Glance: Statewide Time Series. Retrieved April 17, 2018, from http://www.ncdc.noaa.gov/cag/.

are so named because agricultural productivity, ecosystem health, and vegetation processes are most directly affected by soil moisture availability. Typically, these deficits follow extended periods of meteorological drought, but other processes can also be important. Abnormally high temperatures, for example, can amplify soil moisture losses by increasing evaporation. Various characteristics of the land surface (e.g., soil texture and land cover) can also affect the rate and magnitude of changes in soil water storage. Because direct long-term observations of soil moisture are often not available, agricultural drought is frequently monitored using indices calculated from more readily available meteorological datasets (as discussed previously). One such index, PDSI, is shown for California from 1896 to 2017 (figure 1.6). Variability in PDSI-based soil moisture closely tracks the precipitation variability shown in figure 1.5 but also shows some significant differences. Notably, the most recent drought period (2012–2016) in this record ranks as the worst (driest) 3-year soil moisture deficit in the record back to 1896, contrasting with the relatively less severe precipitation drought. This additional drying has been attributed to climate change–induced warming trends and increased evaporative losses, leading to the conclusion that climate change significantly

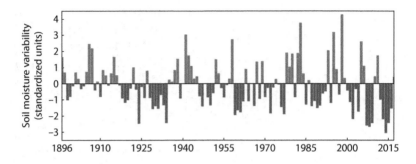

FIGURE 1.6 Summer season (June–August) soil moisture variability over California from 1896 to 2017, represented by the Palmer Drought Severity Index (PDSI). Anomalies represent wetter (positive) or drier (negative) conditions relative to the long-term 1901–2000 baseline. Because long-term warming over California has amplified surface drying (by increasing evaporative losses), recent drought years qualify as some of the most severe in the historical record. *Source:* Updated version of data used in Williams et al., 2015, provided by A. Park Williams.

amplified the soil moisture deficits (Griffin & Anchukaitis, 2014; Williams et al., 2015). Compared to conclusions based on precipitation data alone, this highlights the diversity of processes influencing drought dynamics and their importance when investigating the influence of climate change.

If the precipitation and soil moisture deficits continue to accumulate over time, a drought event is likely to propagate further across the hydrologic cycle, causing declines in water storage in groundwater aquifers, lakes, rivers, and surface reservoirs. Such a scenario is termed a *hydrological drought*, and, as with agricultural droughts, it is sensitive to losses from evapotranspiration and processes beyond simply precipitation. The social, economic, and ecological impacts of hydrologic droughts can be far reaching, especially given the dependence of wildlife, ecosystems, agriculture, and municipalities on the moisture supplied by surface and belowground reservoirs. For California, the hydrologic impact of three multiyear droughts in the past two decades can be clearly seen in observations from the Gravity Recovery and Climate Experiment (GRACE) satellites. These satellites record variations in the gravitational pull of any given location on Earth, which can then be used to estimate changes in total terrestrial water storage, including soil moisture, natural and artificial surface reservoirs, and groundwater. From GRACE satellite images, the impact of the three recent waves of drought on

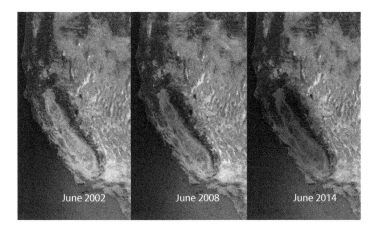

FIGURE 1.7 Cumulative changes in terrestrial water storage over California from 2002 to 2014, as measured from the Gravity Recovery and Climate Experiment (GRACE) satellites. It is estimated that the state lost ~11 trillion gallons of water due to consecutive drought events over this interval, equivalent to ~1.5 times the maximum capacity of Lake Mead, the largest surface reservoir in the western United States. This estimate includes changes in water storage at the surface (lakes, rivers, and snow), as well as declines in soil moisture and groundwater. *Source:* NASA/JPL-Caltech/University of California, Irvine, https://photojournal.jpl.nasa.gov/catalog/PIA18816.

total terrestrial water storage in California is clear (figure 1.7). From 2000 to 2014, California lost an estimated 11 trillion gallons of water, an amount equivalent to 1.5 times the capacity of Lake Mead. This estimate includes not only the water lost because of the droughts themselves but also the groundwater pumped by people in the state to mitigate the drought impacts. This is a clear illustration of how human activities can interact with drought, amplifying hydrologic drought by exploiting groundwater to mitigate drought impacts.

Drought Impacts

Droughts can have far-reaching negative consequences for human and natural systems, the extent of which depends on the characteristics of the drought itself (e.g., duration and intensity) and the vulnerability of the systems in question. Such impacts can include (but are not limited to) reductions in agricultural yields, wildfires, reduced vegetation productivity, increased vegetation mortality, municipal water shortages, inhibited access to water courses for navigation and recreation, and losses of ecosystem services. The impacts of the most recent drought in California have been wide reaching, and it is difficult to find a location, system, or population in the state that has not been affected, to some degree, by this drought.

By the fall of 2016, the 5 preceding years of drought were estimated to have killed over 100 million trees in the state (Stevens, 2016). Losses of these trees represent important changes to habitats, local carbon and water cycles, and a variety of ecosystem services. For people, the most direct and clearest impact of the drought was through agriculture. At its peak in 2015, the drought cost the state's agricultural sector an estimated $2.7 billion and 21,000 jobs (Howitt et al, 2015). The following year the drought eased somewhat, but still caused $600 million in agricultural losses (Medellín-Azuara et al., 2016). Across the state, water prices increased in response to the drought, even in locations where effective conservation eased shortages because the fixed costs required to maintain water infrastructure still needed to be met (McPhate, 2017). The lack of groundwater recharge during the drought caused wells to dry up, leaving some communities in the Central Valley without running water (Lurie, 2015; Siegler, 2017).

Other impacts of the drought were less intuitive—but no less important. From 2011 to 2014, declines in reservoir storage resulted in a 60 percent decline in hydropower production (Hardin et al., 2017), forcing the state to generate more power from fossil fuel sources (primarily natural gas) and increasing electricity costs by $2 billion (Gleick, 2017). The drought caused an increase in cases of West Nile virus as natural sources of freshwater began to dry up and disappear (Palermo, 2015). This caused increased contact between people and mosquitos (e.g., mosquitos began to concentrate around artificial water sources like pools) and between mosquitos and amplifier hosts (birds) that incubate and help spread the disease (because both concentrate around what few water sources are available), as well as changes in mosquito behavior that increased the likelihood of disease transmittance. The drought may have also caused increases in property (but not violent) crimes, likely because increases in water rates hit lower-income families the hardest (Goin et al., 2017).

TWO

Global Hydroclimatology

A lthough Earth's total water budget is effectively closed (with no appreciable losses or additions from one year to the next), water resources are not evenly distributed throughout the climate system in space or time. The movement and distribution of water in all its phases is intimately connected with the Earth's climate, and the study of this interaction between the water cycle and the climate system is called *hydroclimatology*. This chapter presents a broad survey of the processes in the Earth's climate system that control the geographic and seasonal variations of moisture availability. Where on Earth are the most arid regions found, and why? What drives the summer monsoons in Asia, West Africa, and other regions? How does the local climate influence the severity and impact of droughts? And how much of the world's population is regularly exposed to moisture stress? We begin with a brief, but necessary, overview of the global climate system.

The Global Energy Balance

The Earth's climate system involves the cycling and transformation of energy and materials (including water) through the five major components of the Earth system: the atmosphere, the global ocean, the land surface, the biosphere, and the geosphere (in this book, we focus on the first four spheres).

The fundamental source of energy for all climate processes is the sun, which provides the Earth System with ~340 W m^{-2} of radiant energy (averaged globally and over the year), arriving primarily in the form of shorter, visible wavelengths (~0.4 to 0.7 μm). Not all of this energy is absorbed and used to do the "work" of the climate system, however. A fraction (~29 percent) is reflected to space, defined as the Earth's albedo, much of which is controlled by water in various phases. Most of this total is reflected by clouds, which are composed of liquid water and solid ice. At the surface, ice and snow have the highest albedos, reflecting up to 90 percent of local incident solar radiation. Energy reflected by the Earth's albedo is lost and unused by the climate system. Most of the remaining solar energy (~47 percent) is absorbed by the surface, and a smaller fraction (~24 percent) is absorbed by the atmosphere.

The absorption of solar energy at the surface creates a relative surplus of energy at the surface relative to the atmosphere, energy that is transferred to the atmosphere in three ways. The first is through the emission of long-wave or thermal radiation (infrared wavelengths, ~0.7 μm to 1 mm), some of which escapes out to space and some of which is absorbed by greenhouse gases in the atmosphere, including carbon dioxide (CO_2), methane (CH_4), and water vapor. The energy absorbed by these greenhouse gases heats the atmosphere, and, as the atmosphere warms, some of this energy is reradiated back to the surface, providing additional heating beyond what would be available from solar energy inputs alone. Water vapor accounts for about half of the greenhouse gas effect, with clouds separately contributing about 30 percent to the longwave trapping capacity of the atmosphere. This **greenhouse effect** is critical for making our planet habitable, resulting in surface temperatures ~33°C warmer than they would be otherwise. Human activities (primarily land-use change and fossil fuel burning) are increasing concentrations of greenhouse gases (primarily CO_2) in the atmosphere. This is causing more of the outgoing longwave radiation to be absorbed by the atmosphere, increasing global temperatures and affecting the climate system in a variety of ways. Anthropogenic climate change, and its effect on the hydroclimate, will be addressed in chapter 5.

Energy can also be transferred from the surface to the atmosphere through the heating of the air in direct contact with the warmer surface, referred to as *sensible heating*. This creates warmer parcels of air that rise and cool, further redistributing heat and energy through the atmosphere via convection. Thermals (small-scale rising currents of air often exploited by birds and unpowered aircraft) are an example of sensible heating. The last mode of energy transfer, and the most important from a hydroclimate perspective, is *latent heating*, referring to the energy used to evaporate and transpire water from the surface. Evapotranspiration is an endothermic process, absorbing large quantities of energy (~2.5 MJ kg^{-1}) from the surrounding environment (the land and ocean surfaces). About 25 percent of incoming solar energy (~55 percent of solar energy absorbed by the surface) is transferred back to the atmosphere through latent heating. This water, now in the vapor phase, can be entrained in rising air parcels that cool and condense to form clouds and precipitation, releasing this energy back to the environment (now the atmosphere). Latent heating thus represents a critical link between water and energy in the climate system, and, because of this, the hydrologic cycle (especially precipitation and evapotranspiration) is intimately connected to climate system processes.

Seasonal and Latitudinal Distribution of Energy

Energy from the sun is not evenly distributed across the planet or over the course of the year because of (1) the shape of the Earth and (2) the tilt of the Earth's axis. The Earth is approximately a sphere (actually an oblate spheroid, slightly squashed at the poles and bulging along the equator). Because of this shape, energy from the sun hits the equator at a more direct angle (i.e., closer to perpendicular) than at higher latitudes. The more direct the angle, the greater the amount of energy concentrated over a smaller area, resulting in a higher energy flux or density (energy per unit area). As one travels away from the equator (either north or south), the angle of incidence becomes increasingly shallow and more indirect, and the same amount of energy is spread over a much larger area. Integrated over the year, the amount of incoming solar energy progressively declines as one moves from the equator toward the poles.

The tilt (or obliquity) of the Earth's axis is primarily responsible for seasonal differences in this incoming solar energy. Earth's axis of rotation is currently tilted at an angle of ~23.5° relative to the plane of revolution with the sun (the angle of tilt and its orientation with respect to the sun change over timescales of thousands of years). This causes, at various times of year, either the Northern or the Southern Hemisphere to be preferentially tilted toward or away from the sun. The hemisphere tilted toward the sun receives the sun's energy at a more direct angle, increasing the energy flux and experiencing summer. Conversely, the hemisphere tilted away from the sun receives solar energy at a much more indirect angle, experiencing winter. The summer-to-winter difference in energy flux induced by the tilt varies by latitude and is smallest at the equator and largest at the poles, which experience 6 months of daylight during summer (when the sun never drops below the horizon) and 6 months of darkness during winter (when the sun never rises above the horizon). Through the combined influence of the Earth's shape and tilt, there is a progressive decrease in total incoming solar energy and an increase in seasonality as one travels across latitudes from equator to pole.

The General Circulation

Averaged over the year at the top of the atmosphere, there is a net surplus of energy (net solar energy [incoming minus reflected solar energy] > outgoing longwave energy) at the equator and in the tropics and a net energy deficit (net solar energy < outgoing longwave energy) at higher latitudes and the poles. These differences create gradients in energy that generate circulations in the atmosphere that transport energy to offset these imbalances. These circulations, in turn, have major impacts on the seasonality and distribution of precipitation, water resources, and climate zones throughout the world.

The general circulation in the atmosphere begins with solar heating in the tropics. Heating at the surface near the equator creates a region of low pressure near the surface and rising motion in the atmosphere, called the Intertropical Convergence Zone (ITCZ). Heating along the ITCZ evaporates water from the surface, providing energy for convection, leading to cloud formation and some of the highest precipitation rates in the world. At upper levels in the atmosphere, this air moves horizontally toward the poles, eventually subsiding (sinking) at approximately 30°N and 30°S latitude. This sinking motion suppresses convection and cloud formation, creating zonal bands of semipermanent high pressure near the surface. Flow near the surface returns to the equator as the northeast and southeast trade winds (the names of these and most winds refer to the direction from which the winds originate; i.e., northeast winds flow from a northeasterly direction), deflected by the Coriolis force to the right in the Northern Hemisphere and to the left in the Southern Hemisphere. This circulation, characterized by rising motion in the tropics and sinking motion in the subtropics, is called the **Hadley circulation**. Poleward flow from the subtropical latitudes is similarly deflected by Coriolis, creating surface westerlies in both hemispheres. This relatively warm, subtropical air encounters cooler air from higher latitudes, creating a secondary region of instability, rising motion, and enhanced precipitation in the midlatitudes. In this region, energy transport from lower latitudes toward the poles is dominated by midlatitude storms and frontal systems. Higher in the atmosphere, both in the subtropics and at higher latitudes, meridional gradients (gradients across latitudes) in temperature result in concentrated regions of air moving rapidly from west to east known as the jet streams. The jet streams are dynamic, changing shape and speed, steering regional weather systems, and acting as permeable barriers between warmer air from low latitudes and colder air from high latitudes. The strength of the jet streams is directly related to the meridional temperature gradient at upper levels in the atmosphere and is stronger during winter, when these gradients are steeper.

This simplified, zonally symmetric picture of the general circulation is broadly correct, but it is further complicated by two factors: (1) the seasonal migration of the latitude of maximum solar heating and (2) the heterogeneous distribution of land and ocean. Because of the tilt of the Earth, the latitude of maximum solar heating varies across the year, reaching a maximum latitude of ~23.5° in the summer hemisphere. As this zone of maximum heating (known as the *thermal equator*) moves across latitudes, the ITCZ shifts, and many other features of the general circulation (e.g., the Hadley cells and jet streams) follow along. Beyond zonal and seasonal differences in heating and energy, the land and ocean respond differently to heating because of differences in their *heat capacity*. A unit mass of ocean (or any body of water) requires about four times the energy of an equivalent mass of land to increase its temperature by one degree. Land areas will therefore respond much more rapidly to changes in energy inputs, heating up more quickly relative to the ocean during the summer (or daytime) and cooling more quickly during the winter (or nighttime). These differences in temperature between land and ocean can affect local and regional climates (temperature and precipitation), as well as energy and pressure gradients. Combined, these seasonal shifts and differences in heat capacity between land and ocean act to disrupt the annual average and zonally symmetric circulation, creating a variety of diverse climate regimes across the world.

Humidity, Precipitation, and the General Circulation

The general concept of **humidity** refers to some measure of water vapor content in the air, either the actual amount of water vapor present or the amount present relative to the saturation capacity. The former can be expressed in a variety of ways, including the *specific humidity* (the ratio of the mass of water vapor to the total mass of air), the *mixing ratio* (the ratio of the mass of water vapor to the mass of dry air), or the *vapor pressure* (the pressure exerted by the water vapor molecules). The amount present relative to the saturation capacity is the *relative humidity*, with the saturation capacity determined primarily by temperature via the Clausius-Clapeyron relationship described in chapter 1. Water vapor in the air condenses to form clouds when the temperature cools to the *dew point* and the air reaches supersaturation (over 100 percent relative humidity). Once this happens, precipitation occurs if these cloud droplets aggregate and become large enough to fall to the surface without reevaporating.

Air most commonly becomes supersaturated through adiabatic cooling. As parcels of air rise in the atmosphere, they expand and cool, even without exchanges of energy between the parcels and the surrounding free atmosphere. If these parcels cool enough to reach the dew point, condensation will begin, and a cloud will form. This rising motion often occurs in response to direct heating of the surface, which creates convective instabilities resulting in rising air parcels. Thunderstorms and cumulonimbus clouds, common in the tropics and during the summer, are examples of such dynamics. In midlatitude regions (often during the winter or cold season), the ascent and associated cooling often occurs along a front, the transition zone between two air masses of different temperatures. Because warm air is less dense and more buoyant than cold air, the warm air mass is forced to rise over the cooler air. Near the surface, supersaturation can occur when radiative cooling takes place or when air masses lose heat as they pass over cold surfaces; in such scenarios, the result is fog. Clouds may also form if air is forced to ascend because of topography, such as when prevailing winds encounter a large mountain range. Rising motion, and therefore precipitation, is thus favored by low pressure and mass convergence near the surface. Conversely, high pressure near the surface is associated with sinking motion and mass divergence near the surface because sinking air warms adiabatically, decreasing relative humidity and making it difficult for parcels to reach supersaturation. Regions of high pressure near the surface therefore typically have low precipitation. Beyond the necessary cooling to reach supersaturation, the other major requirement for precipitation is a source of moisture. Globally, the ocean (~70 percent of the global surface area) is the dominant source for this moisture, although more localized moisture recycling is common in some land regions. By some estimates, for example, ~25 percent of precipitation in the Amazon originates as local evapotranspiration from the land surface (Eltahir & Bras, 1994).

Precipitation occurrence and intensity are intimately connected with circulation in the atmosphere, modulated by both moisture availability and circulations that facilitate the vertical movement of air parcels. The importance of these circulations, and their seasonal migrations, can be clearly seen when examining global maps of seasonal precipitation (figure 2.1). During boreal summer (in the Northern Hemisphere; June–July–August, or JJA), the ITCZ follows the thermal equator northward, shifting the ascending branch of the Hadley circulation and associated region of high precipitation into the Northern Hemisphere. The pole-to-equator temperature gradient weakens, slowing the jet stream and causing a poleward migration of the storm tracks. Other major global circulation features follow suit, shifting the regions of moisture convergence and divergence. During boreal winter (December–January–February, or DJF), the thermal equator migrates back to the Southern Hemisphere, and other features of the general circulation similarly shift. These shifts in circulation explain the dramatic seasonality of precipitation in many areas, such as the relatively dry summers and wet winters in

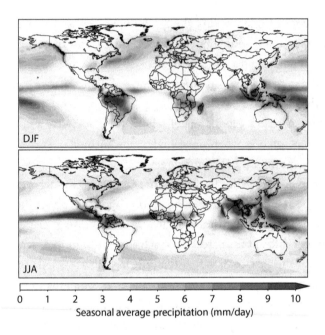

FIGURE 2.1 Global maps of seasonal average precipitation for boreal winter (December–January–February, or DJF) and boreal summer (June–July–August, or JJA) from 1979 to 2017. The geographic distribution and seasonal shifts in precipitation are broadly controlled by the general circulation of the atmosphere. For example, the northward shift in tropical precipitation during JJA closely follows the seasonal migration of the thermal equator and Intertropical Convergence Zone. *Source*: Data from Global Precipitation Climatology Project, National Center for Environmental Information, https://www.ncei.noaa.gov/data/global-precipitation-climatology-project-gpcp-monthly/access/.

California and the reverse precipitation seasonality over monsoon regions like India. In other regions, such as the eastern United States, precipitation is more evenly distributed across the year. However, even in these places, the character of precipitation may change substantially across the year, shifting, for example, from snow and large-scale frontal systems in the winter to rain and more localized convective activity in the summer.

Hydroclimate and Aridity

As discussed in chapter 1, *aridity* is a permanent or semipermanent state of the climate typically associated with low precipitation. Hydroclimate and aridity, however, are not solely determined by precipitation, and it is important to use more integrated metrics of the surface water balance that account for both moisture inputs (e.g., precipitation) and losses (e.g., evapotranspiration). One simple but useful way to represent this is simply through the difference of evapotranspiration minus precipitation $(E - P)$ (figure 2.2). In the time average, $E - P$ must be balanced by moisture divergence. In the tropics and higher latitudes, $E - P$ is generally negative (greenish-blue shading). These are regions of net moisture convergence, generally associated with rising atmospheric motions, abundant water available at the surface, and high precipitation rates. In the subtropics, $E - P$ is positive (brown shading), indicating net moisture divergence in these latitude bands that sit beneath the descending branches of the Hadley circulation. Even this zonal picture, however, is broken up by the distribution of land and ocean, which, among other things, can impart significant seasonality in the hydroclimate. For example, much of Southeast Asia, India, and the Himalayas are regions of negative $E - P$ (net moisture convergence) in the annual mean, despite the fact that similar latitudes over the ocean have positive $E - P$. These land regions lie in the heart of the Asian monsoon (see below), a phenomenon that brings in substantial atmospheric moisture and rain during the boreal summer, which more than compensates for the extended winter dry season.

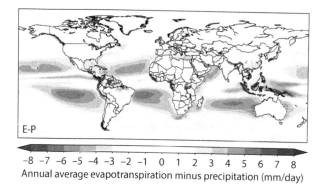

E-P

-8 -7 -6 -5 -4 -3 -2 -1   0   1   2   3   4   5   6   7   8
Annual average evapotranspiration minus precipitation (mm/day)

FIGURE 2.2 Annual average evapotranspiration minus precipitation $(E - P)$. Regions of negative $E - P$, indicating net moisture convergence, are widespread across the tropics and middle to high latitudes and are associated with ample precipitation and generally humid climates. The most widespread regions of positive $E - P$, indicating net moisture divergence and more arid climate regions, are centered in the subtropics and associated with the descending branches of the Hadley circulation. *Source:* Data from MERRA-2 Reanalysis. The data used in this effort were acquired as part of the activities of NASA's Science Mission Directorate and are archived and distributed by the Goddard Earth Sciences (GES) Data and Information Services Center (DISC), http://apdrc.soest.hawaii.edu/datadoc/merra2.php.

TABLE 2.1
Classification of Drylands Areas Based on Mean Annual Precipitation and the
Aridity Index

Climate Zone	Mean Annual Precipitation (mm/year)	Aridity Index
Hyperarid	<100	<0.05
Arid	100–250	0.05–0.20
Semiarid	250–600	0.20–0.50
Dry subhumid	600–1,200	0.50–0.65

Alternatively, hydroclimate can be classified using potential evapotranspiration (PET), the amount of water that would evapotranspire from the surface to satisfy atmospheric demand given unlimited water availability. A common way to express this is through the aridity index (AI), defined as the ratio of precipitation to potential evapotranspiration ($AI = P/PET$). The AI is widely used to classify arid and semiarid environments over land (table 2.1, figure 2.3) and yields distributions of hydroclimate that often differ substantially from the patterns of $E - P$ shown previously. The driest categories in the AI capture

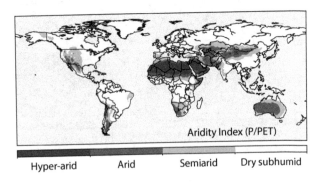

Aridity Index (P/PET)

Hyper-arid   Arid   Semiarid   Dry subhumid

FIGURE 2.3 The global distribution of aridity zones, calculated using the aridity index ($AI = P/PET$), using climate data from 1951 to 2000 (see table 2.1). The most widespread arid and hyperarid regions over land are clustered in the subtropics and associated with some of the largest deserts in the world, including the Sahara and Arabian Deserts. Many of the largest savanna and grassland ecosystems occur in semiarid and dry subhumid regions, including the central plains of North America and the Sahel region of Africa. *Source*: Data from the Climate Research Unit of the University of East Anglia, high-resolution gridded datasets, version 4.01, https://crudata.uea.ac.uk/cru/data/hrg/.

the major desert regions of the world, including the Sahara, the Arabian Peninsula, and the Taklamakan Desert in Central Asia. The semiarid and dry subhumid classifications encompass the major grassland and savanna regions, areas where grasses and shrubs dominate and conditions are too dry for closed canopy forests. As with $E - P$, however, this is a broad and fairly simple classification that glosses over important seasonal dynamics (e.g., seasonality of precipitation or temperature) in many of these regions.

Mediterranean Climate

Mediterranean climate regions (figure 2.4) occur primarily along the western coasts of continents, areas sitting on the eastward flanks of the five semipermanent subtropical high-pressure cells over the oceans. These high-pressure areas (anticyclones) are associated with the descending branches of the Hadley circulation, so Mediterranean climates are generally found between 30° and 45° latitude in both hemispheres. Mediterranean climates are characterized by cool, wet winters and warm, dry summers, a pattern caused by the seasonal migration of these anticyclones. In the summer, these subtropical anticyclones intensify and expand, dominating regions with a Mediterranean climate. Subsidence prevails, storms are deflected away from the coast, and precipitation is suppressed, causing an extended dry season during the summer. During the winter or cold season, the high-pressure centers move

FIGURE 2.4 The global distribution of regions that have a Mediterranean climate. These areas are centered in the midlatitudes of both hemispheres, situated predominantly along the western coasts of continents. They are characterized by cool (but typically not freezing) winters with significant precipitation and by warm or hot summers with little or no precipitation and often with high levels of evaporative demand. *Source*: Data from the Climate Research Unit of the University of East Anglia, high-resolution gridded datasets, version 4.01, https://crudata.uea.ac.uk/cru/data/hrg/.

equatorward as they follow the seasonal migration of the thermal equator and general circulation. The prevailing winds shift in response to favor movement of moisture from the oceans onto land and allow penetration of storm tracks into these regions, making precipitation likely during these months.

The annual amplitude of temperature in Mediterranean climate regions is relatively low compared to precipitation, with winter temperatures strongly moderated by the advection of air from the oceans. Summer temperatures away from the coast and the moderating influence of the ocean, however, can be extreme. Because of the out-of-phase peaks in precipitation and tempera-ture, areas of Mediterranean climate typically experience peak moisture stress (low precipitation and high temperatures) during the summer dry season. As the name implies, most of the regions bordering the Mediterranean Sea have a Mediterranean climate, as do California, southwestern Australia, central Chile, and parts of coastal South Africa (figure 2.5). Because of the

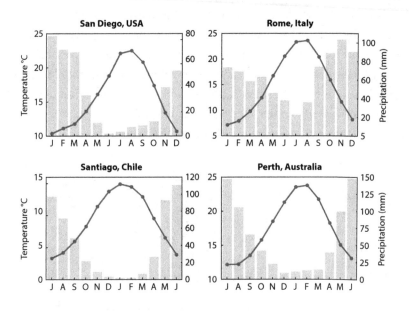

FIGURE 2.5 Average seasonal cycles in temperature (lines) and precipitation (bars) from 1951 to 2000 for four cities that have a Mediterranean climate: San Diego, United States; Rome, Italy; Santiago, Chile; and Perth, Australia. Although some differences are apparent, all four cities share the same major features of a Mediterranean climate, including a winter season peak in precipitation and an extended summer dry season. *Source:* Data from the Climate Research Unit of the University of East Anglia, high-resolution gridded datasets, version 4.01, https://crudata.uea.ac.uk/cru/data/hrg/.

extended warm dry seasons, vegetation in Mediterranean climate regions is typically adapted to drought and aridity, with local ecosystems dominated by shrubs, grasses, and open woodlands.

Monsoon Climate

Monsoon climate regions feature wet summers and dry winters, superficially appearing as the inverse of Mediterranean climate regions. The underlying dynamics, however, are quite different. Monsoon climates occur primarily in the tropics and subtropics and are defined by pronounced seasonal reversals in wind speed and direction. During summer, intense solar heating warms the land surface, reducing surface pressure. This drives low-level convergence and advection of moisture from neighboring ocean regions, which together lead to intense precipitation over the land during the summer. In the winter, pronounced cooling of the land causes anomalous high pressure at the surface. Wind direction reverses, preferential flow is from the land to the oceans, and precipitation is suppressed over land during this season. This simplified view of monsoon systems is further complicated by other processes that vary across monsoon regions, including topography, the seasonal migration of the ITCZ, and land-atmosphere interactions and feedbacks.

Monsoon regions occur in both hemispheres and on all continents except Antarctica (figure 2.6). The largest and most intensely studied monsoon is the Asian monsoon, a system encompassing India, Southeast Asia, and much of China. Major monsoon systems also occur in South America, southern Africa, northern Australia, West Africa, and southwestern North America. In July, the latitude of maximum insolation migrates to the Northern Hemisphere, causing the heating of land areas and the development of southerly and southwesterly offshore winds that bring moisture into Central America, West Africa, and Asia. In January, the thermal equator migrates south, and the winds in these same regions reverse. In the Southern Hemisphere, heating over northern Australia, southern Africa, and South America in austral summer causes similar wind reversals that bring moisture and rainfall into these regions. As with Mediterranean climate regions, the seasonal amplitude of temperature in most monsoon regions is relatively low compared to the much more extreme seasonality of precipitation (figure 2.7).

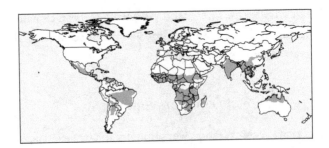

FIGURE 2.6 The global distribution of major areas that have a monsoon climate. These areas are concentrated primarily in the tropics and subtropics. Precipitation is intensely seasonal, with the vast majority falling during the summer. One of the largest continuous monsoon area encompasses India, Southeast Asia, and parts of China. Other important regions with a monsoon climate include West Africa, North America, Australia, South America, and South Africa. *Source*: Data from the Climate Research Unit of the University of East Anglia, high-resolution gridded datasets, version 4.01, https://crudata.uea.ac.uk /cru/data/hrg/.

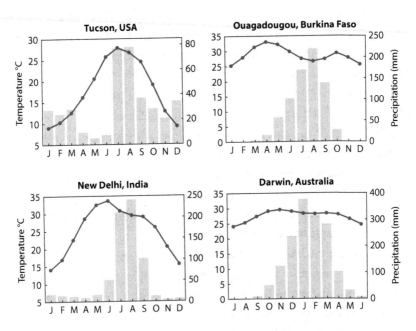

FIGURE 2.7 Average seasonal cycles in temperature (lines) and precipitation (bars) from 1951 to 2000 for four cities that have a monsoon climate: Tucson, United States; Ouagadougou, Burkina Faso; New Delhi, India; and Darwin, Australia. All four cities share the same major features of a monsoon climate, including a summer season peak in precipitation and an extended winter dry season. *Source*: Data from the Climate Research Unit of the University of East Anglia, high-resolution gridded datasets, version 4.01, https://crudata.uea.ac.uk/cru/data/hrg/.

Snow

Where winter temperatures are consistently below freezing (mountainous areas and higher latitudes), a significant fraction of precipitation during these months falls as snow. In the coldest of these regions, temperatures may not even get warm enough to melt snow until the spring. In such cases, the surface snowpack accumulates over the course of the winter before rapidly melting over several weeks in the spring. The result is a large runoff pulse in these areas that rapidly recharges lakes, rivers, and soil moisture at the surface just prior to the summer season, with its high evaporative demand. In such regions, both human and natural activities have adapted to work around the unique dynamics created by this snowmelt pulse.

The extent of global snow cover peaks during boreal winter, a consequence of the much greater extent of high-latitude land areas in the Northern Hemisphere. Accumulation begins in the fall, with the snowpack reaching a typical annual maximum in January or February. The snowpack melts rapidly with the onset of spring, followed by the annual minimum during boreal summer. The timing of snowmelt varies from year to year and spatially, with earlier melt occurring during anomalously warm years and in warmer climates at lower elevations and latitudes. Spring season peaks in discharge and streamflow occur in many major river systems in response to the spring snowmelt pulse. These include the Colorado and Rio Grande in North America, the Mekong and Indus in Asia, and the Fraser and Mackenzie in Canada. Streamflow in the smaller but still regionally important rivers like the Sacramento in California and the Columbia in Washington and Oregon is also snowmelt driven. Snowmelt can even influence the flow in river systems that sit primarily in the tropics or in warmer, snow-free regions. Most of the Mekong River Basin, for example, sits squarely within the tropical monsoon climate area of Southeast Asia. The Mekong's upper basin, however, is on the Tibetan Plateau, where spring snowmelt provides a significant proportion of the total annual runoff and streamflow. Similarly, snowmelt in Colorado provides runoff to the Rio Grande and Colorado Rivers, systems that eventually make their way to warmer latitudes in the southwestern United States and Mexico. Lack of snowmelt in the spring, either because the snow has melted early or a warm winter has caused a larger fraction of precipitation to fall as rain, can lead to drought conditions in the summer, even if total precipitation is near-normal.

Arid Regions

Arid regions, often referred to as **deserts**, occur where potential evapo-transpiration far exceeds precipitation (i.e., where we find the most extreme values of the AI; see figure 2.3). They occupy many areas across latitudes and in both hemispheres. Vegetation in arid regions typically has low density and productivity and is well adapted to extended periods of low moisture availability. As such, deserts are readily identifiable from satellite data that show low levels of vegetation activity or productivity throughout the year (figure 2.8). Deserts occur where prevailing conditions inhibit precipitation, either because the dominant circulation makes it difficult for air parcels to rise and cool to supersaturation (thereby suppressing cloud formation and precipitation) or because atmospheric moisture availability is low.

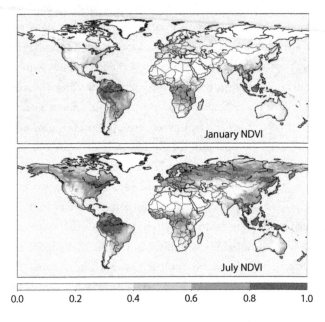

FIGURE 2.8 Mean January and July values from the Normalized Difference Vegetation Index (NDVI), a measure of plant cover and productivity, for 1982–2015. Low vegetation activity throughout the year characterizes the major desert regions, including the Sahara in northern Africa, the Simpson in Australia, the Great Basin in western North American, the Atacama in Chile, and the Taklamakan in Central Asia. *Source:* Data from GIMMS AVHRR Global NDVI Dataset, version 3g.v1, University of Maryland, https:// ecocast.arc.nasa.gov/data/pub/gimms.

The most extensive deserts on Earth are found in the subtropics, where subsidence from the descending branches of the Hadley circulation inhibits convection and causes a net divergence of atmospheric moisture. These latitude bands contain some of the largest desert regions in the world, including the Sahara in northern Africa and the deserts in the interior of Australia. Deserts also commonly occur on the leeward sides of mountain ranges. When prevailing winds encounter mountains, this air is forced to rise and cool. If the air contains sufficient moisture, this cooling will cause clouds to form and precipitation to occur on the windward side. By the time the air crosses over the mountain range, it will have lost significant moisture. Further, as it descends on the leeward side, the air will warm, lowering relative humidity. Combined, these processes act to suppress precipitation, creating a "rain shadow" desert on the leeward side. The Great Basin in the interior west of North America is a desert created by such conditions, as prevailing westerly winds from the Pacific Ocean encounter the Coastal Range and Sierra Nevada in California. These mountains force precipitation out along the coast and in California, leaving little moisture by the time the air reaches Nevada and the Great Basin.

Central Asia is also home to several deserts. Here, mountain ranges and the great distance from the ocean make it difficult for atmospheric moisture to reach this region. In this isolated continental interior, one can find the Taklamakan and Gobi Deserts in Mongolia and western China. Deserts are also common in coastal regions adjacent to intense coastal upwelling, typically found on the western coasts of continents. Upwelling occurs when atmospheric circulation causes surface flow in the ocean away from the coast, allowing much colder water from the ocean depths to come to the surface. As air passes over the cold ocean waters, it cools near the surface as heat is transferred from the atmosphere to the cold ocean. This causes the atmosphere to become more stable, inhibiting convection and precipitation. This mechanism is at least partially responsible for deserts along the Chilean coast in South America, including the Atacama, one of the driest regions in the world. Similar areas occur on the peninsula of Baja California, in the Namib in southern Africa, and in the Atlantic coastal desert in Morocco.

Other Regions

Not all regions fit within these simple hydroclimate categories. Across much of the midlatitudes in both hemispheres (e.g., eastern North America, Europe north of the Mediterranean, and southeastern Australia), precipitation is more equally distributed throughout the year. Winter precipitation often occurs as a result of large-scale frontal systems, and, in colder regions, precipitation during these months may fall predominately as snow rather than rain. During the summer or warm season, precipitation typically arises from much more localized dynamics, often through convection forced by intense solar heating. In the tropics, precipitation seasonality is often associated with the seasonal migration of the ITCZ, resulting in defined wet seasons that are not easily categorized as a monsoon or a Mediterranean climate. In many such regions, however, the "dry" season may still experience substantial precipitation. For example, average precipitation over the year in the city of Manaus, Brazil, situated in the Amazon rainforest and very close to the equator, ranges from a minimum of 50–60 mm in August to over 300 mm in March, with total annual precipitation of over 2,200 mm. By contrast, New York City receives, on average, a little over 1,000 mm of precipitation each year, more or less evenly distributed across all months. Other regions experience still different seasonal patterns of precipitation. In the central plains of the United States, for example, there is a pronounced precipitation peak in spring and early summer. This is caused by the intensification of the Great Plains low-level jet during these months, which brings moisture from the Gulf of Mexico into the interior of North America. Parts of East Africa experience bimodal peaks in precipitation in the boreal spring and fall, a function of complex topography, the seasonal migration of the ITCZ, and the influence of the Indian and Atlantic Oceans. Thus while the global distribution of various hydroclimate regimes is controlled primarily by the general circulation of the atmosphere, a variety of local processes are important for shaping many regional differences.

Hydroclimate Across Time

Variability in the hydroclimate over time in any given region (i.e., droughts and pluvials) is typically small compared to differences across regions with different hydroclimate regimes. A drought in New York, relatively speaking, will still likely be wetter in terms of absolute volumes of water than a normal year in central Arizona. Within a reasonably stationary climate, conditions can generally be expected to revert back to the normal baseline, even after prolonged extreme events.

The current distribution of hydroclimate regimes across the world, however, is not fixed in time. Global and regional climate shifts over the last several thousand to several million years have forced remarkable shifts to new and different hydroclimate regimes that, in some cases, have persisted for hundreds to thousands of years. In such scenarios, the line between transient *droughts* and *pluvials* and permanent changes in *aridity* becomes blurred. Indeed, in a system as dynamic as our climate, temporary versus permanent is often just a matter of timescale. During the mid-Holocene (~9,000 to 6,000 years ago), for example, the West African monsoon was much more vigorous, penetrating far into what is now the Sahara Desert. This wetting of the Sahara, currently one of the driest regions on the planet, was strong enough to support lakes, grassland ecosystems with diverse megafauna, and extensive human habitation. Such a state persisted for several thousand years before another change in global climate caused the relatively abrupt collapse of this extended monsoon and a shift to the more arid Sahara of today. With climate change, many areas are expected to see similar changes in aridity as global warming shifts patterns of precipitation and evaporative demand. Such changes in the future and the past, including the underlying climate system processes involved, will be discussed in chapters 4 and 5.

Connections Among Hydroclimate, Drought, and Water Scarcity

The mean hydroclimate state and precipitation seasonality in a region can significantly affect the manifestation of droughts and their impacts. Precipitation seasonality has little influence on the duration or intensity of meteorological droughts (precipitation deficits), but it can have larger effects as precipitation deficits propagate through the hydrologic cycle into agricultural drought (reduced soil moisture) and hydrological drought (reduced runoff, streamflow, and groundwater) (Van Loon et al., 2014). For regions with significant precipitation in all months, meteorological droughts can cause soil moisture and streamflow deficits to develop, or ameliorate, at any time of the year.

In climates with significant dry seasons, however, there is often strong drought persistence between wet seasons. A wet season that ends in drought is unlikely to recover until the following wet season because little precipitation will fall in the intervening dry months. The drought anomalies and moisture deficits are thus carried forward in time until at least the next wet season, when the seasonal dynamics shift to provide the possibility of enough precipitation to end the drought. Drought persistence is further amplified in regions where the dry season is especially warm because this can exacerbate surface drying by increasing evaporative losses, making even less water available to recharge soil moisture, groundwater, or streamflow.

Such a situation occurs annually in California, where nearly all precipitation falls between October and March. Once the spring and summer dry season begins, any major deficits are unlikely to be ameliorated until the cold season storms arrive the following fall and winter. A somewhat different situation occurs in the larger southwest region of North America. Here, precipitation peaks during the summer, much of it associated with the North American monsoon. However, temperatures and evapotranspiration rates are so high during these months that, in spite of the often-intense thunderstorms supplying precipitation, very little of this water contributes to streamflow or groundwater. Instead, surface recharge occurs mainly in response to precipitation during the winter, when moisture can steadily accumulate because evaporative demand is relatively low. Other issues emerge in regions with significant snowmelt, where hydrologic droughts during the summer are likely to persist at least through the winter because precipitation accumulates in the surface snowpack and is effectively not available until the spring snowmelt pulse.

Droughts generally have larger impacts on people and ecosystems in semiarid and arid regions, where absolute water availability is low. It is therefore important to consider societal and ecosystem water demands and not just climate variability when analyzing drought impacts and risk. Arid and semiarid regions occupy ~30 percent of the global land area, and, by some estimates (Munia et al., 2016; Oki & Kanae, 2006), ~2 to 3 billion people live in these highly water stressed areas. One way to measure or quantify water scarcity in a way that accounts for both physical factors (i.e., supply) and social factors (i.e., demand) is with *baseline water stress*: the ratio of the total annual water withdrawals to the total annual renewable supply available (figure 2.9). Annual renewable supply is typically based on total estimated runoff or streamflow in a region. Water stress peaks during periods when water demands exceed availability (e.g., droughts and dry seasons). Importantly, the water stress indicator in figure 2.9 reflects baseline *average* conditions over a region. During a drought, available water is likely to be much less, in which case total water stress will be much higher. Unsurprisingly, the most water stressed countries are collocated with some of the most arid regions of the world, including northern Africa, the Middle East, Australia, and Central Asia. Water stress can also be high in wetter countries, such as India and China. Despite a normally very wet summer monsoon season, these countries contain large populations (which translate to high water demand and withdrawals) and can experience extended dry seasons (which can mean months of significantly diminished supply). Populations living in areas with high baseline water stress may be especially vulnerable to changes in water scarcity from naturally occurring drought events and climate change (Van Beek et al., 2011).

A further complication when considering water stress and sustainability is that the natural geography of water resources does not cleanly follow

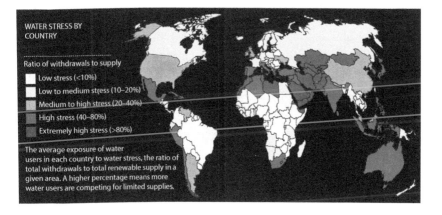

FIGURE 2.9 Baseline average water stress for each country, estimated as the ratio of the total annual water withdrawals to the total annual renewable supply available. Water stress is higher in more arid regions and in areas with high populations and high water demands. *Source:* World Resources Institute, http://www.wri.org.

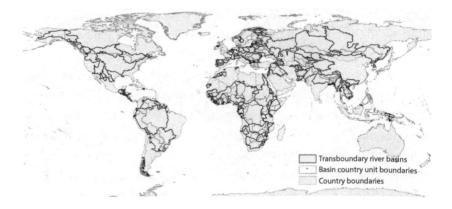

Transboundary river basins
Basin country unit boundaries
Country boundaries

FIGURE 2.10 A global map of major transboundary river basins, which include some of the most important rivers for water resources (e.g., the Colorado, Mekong, and Nile Rivers). Successful management of water resources in these basins requires careful coordination across international boundaries, especially during periods of drought. *Source:* UNEP-DHI and UNEP (2016), Transboundary River Basins: Status and Trends, United Nations Environment Programme (UNEP), Nairobi, http://twap-rivers.org.

political boundaries. Many of the largest river basins, supplying water resources to millions of people, cross international boundaries (figure 2.10). Globally, there are 276 such transboundary lake and river basins, covering almost half the global land surface. These basins account for 60 percent of the global water flows and contain 40 percent of the global population (Munia et al., 2016). Water scarcity in these basins has the potential to create political conflicts, especially between different populations or countries withdrawing water upstream versus downstream in the basin. There is evidence in many transnational basins that upstream water use can significantly exacerbate downstream physical water stress (Munia et al., 2016). But direct causal linkages between water stress and increased conflict are not clear, and there is some evidence that water scarcity can actually foster greater cooperation (Wolf et al., 2003). There is therefore still significant complexity and uncertainty in our understanding of social and political responses to environmental events, such as droughts, and these nonphysical factors are vitally important to consider when assessing impacts, vulnerability, and resiliency.

THREE

Drought in the Climate System

S ignificant droughts of all types occur in nearly every terrestrial region of the planet. Their underlying causes, characteristics, and impacts, however, can differ markedly from region to region and even from one event to the next. Just as there is no single definition of drought, there is no singular mechanism that can explain the occurrence and severity of every drought event. Understanding global drought dynamics therefore requires a broad understanding of how processes and variability connected to the oceans, atmosphere, and land surface all influence moisture supply and demand on a variety of spatiotemporal scales. What processes in the climate system cause droughts to occur and intensify? How well constrained is our understanding of these dynamics, and can we use them to predict droughts in advance? What are the relative roles of precipitation and temperature? And what are some of the consequences for ecosystems when droughts do occur?

Climate, Drought, and People

Most droughts begin as precipitation deficits (figure 1.4), regardless of their ultimate severity and duration, manifestation across the hydrologic cycle, or impact on people and ecosystems. As with the global distribution of terrestrial hydroclimate (discussed in chapter 2), meteorological droughts are often caused by anomalous high pressure in the atmosphere. Anticyclonic circulation associated with this high pressure deflects incoming storms, causes anomalous subsidence and moisture divergence, and suppresses precipitation. Unlike the permanent or semipermanent geographic and seasonal variations in circulation that control the global distribution of hydroclimate regimes, however, droughts and their underlying causes are temporary or transient anomalies relative to the local long-term average conditions. A drought in Florida, for example, represents an anomalously dry departure from average conditions in Florida but may still be wetter (in absolute terms) than a normal year in Arizona. Droughts can occur anywhere from the deserts and semiarid grasslands of western North America to the rainforests of the Amazon, arise through a variety of mechanisms, and have a broad range of impacts.

Behind the circulations that directly cause most meteorological droughts are the ultimate drivers of these circulation anomalies themselves, as well as other processes that amplify moisture deficits across the hydrologic cycle (e.g., soil moisture and streamflow). These drivers include changes in ocean circulation, feedbacks from interactions between the atmosphere and the land surface, and even random variability within the atmosphere itself. Many of these are connected to major climate modes, features of the ocean and atmosphere system that have preferred, and often cyclically recurring, patterns of temporal and spatial variability. Through the influence of these modes, droughts can be expected to occur with some regularity and often some degree of predictability in many regions (e.g., southwestern North America). Not all regions, however, have such predictable or regular drought dynamics, making predicting and preparing for such events in advance significantly challenging. Human activities also have direct and indirect influences on physical drought dynamics and impacts on the climate system. Increased greenhouse gas concentrations from deforestation and the burning of fossil fuels affect the global climate, influencing processes that control moisture supply and demand and very likely causing significant changes in drought and hydroclimate in the coming decades. Locally, land-use changes, including deforestation and urbanization, can affect the water retention capacity of soils, partitioning of the water and energy balance at the surface, and interactions between the surface and the atmosphere. Importantly, human impacts on drought (e.g., through land-use changes or climate change) mostly act through existing pathways in the Earth system rather than through the introduction of new or novel mechanisms. For example, greenhouse warming may amplify evaporative demand, increasing evaporative losses from the surface, and increases in impermeable surfaces from urbanization can decrease infiltration of water into the soil and increase runoff.

Modern and historical societies have developed numerous strategies to ameliorate the effects of droughts when they do occur and facilitate the expansion of people into regions previously too arid to support large human populations and agriculture. The building of dams and reservoirs, as well as increased exploitation of groundwater, redistributes existing water resources and increases availability for all human water needs. Irrigation, drawn from natural and artificial reservoirs, supplies moisture to agricultural fields, compensating for precipitation deficits during droughts and permanent lack of rainfall in more arid regions. Climate and societies, however, are not static. Populations and water demands can grow, climate can shift to drier states for often prolonged periods of time, and previously available resources from groundwater or surface reservoirs may be eventually depleted. All these factors can contribute to the manifestation and impacts of drought within a given region. Analyses of drought and sustainability therefore need to address the constantly shifting balance of moisture supply and demand and the degree to which they are each dependent on the physical climate system and sociopolitical factors.

Teleconnections Between Regional Hydroclimate and Ocean Variability

Interactions between the ocean and the atmosphere are especially strong in the tropics. Variations in tropical ocean temperatures from the Atlantic, Pacific, and Indian Ocean basins influence circulation patterns in the atmosphere locally, and these changes can propagate through the atmosphere for thousands of kilometers. In this way, the atmosphere acts as a bridge that creates **teleconnections** between the tropical oceans and the climate in distant regions throughout the world. In regions where such teleconnections give rise to storm track changes, anomalous subsidence or rising motion, or changes in moisture convergence, precipitation will respond in turn, potentially giving rise to floods, droughts, and pluvials.

The **El Niño Southern Oscillation (ENSO)** is the most important and best understood mode of climate variability in the world, acting as the dominant driver of interannual to decadal drought variability in many regions

(C. Wang et al., 2017). ENSO itself is a phenomenon arising from coupled interactions between the ocean and atmosphere in the equatorial Pacific (figure 3.1). In the normal, or neutral, ENSO state, strong heating over the Maritime Continent (islands off the coast of Southeast Asia, including Indonesia, the Philippines, and Papua New Guinea) and the Indo-Pacific Warm Pool (a region of exceptionally warm water in the shallow seas surrounding the Maritime Continent; sea surface temperatures here exceed 28°C) drives vigorous convection, ascending motion in the atmosphere, and high levels of precipitation over this region. Aloft, this air travels eastward, eventually subsiding over the much cooler waters of the eastern tropical Pacific before traveling back to the western tropical Pacific near the surface as the trade winds. This zonal (east-west) circulation in the atmosphere is referred to as the *Walker circulation*. The trade winds drive upwelling of deep, cold ocean water in the eastern tropical Pacific and push warm surface water toward the Maritime Continent, enhancing the warm pool. This zonal gradient in sea surface temperatures, with warmer water in the west and cooler water in the east, amplifies the surface pressure gradients that drive the Walker cell, feeding back to reinforce the circulation in the atmosphere.

Every few years, however, the trade winds weaken, reducing upwelling in the east and allowing warm surface water in the west to move toward the central and eastern tropical Pacific. This weakening of the sea surface temperature gradient further reduces the zonal pressure gradient in the atmosphere, weakening the Walker circulation and trade winds. The region of dominant heating, ascent, and convection shifts eastward from the Maritime Continent toward the central and eastern tropical Pacific. This pushes the system into an El Niño event, or warm phase, characterized by weaker trade winds, reduced upwelling in the eastern tropical Pacific, anomalous subsidence over the Maritime Continent, and enhanced convection and precipitation over the central and eastern tropical Pacific. In the months after an El Niño event peaks (usually November or December), the system typically shifts toward more neutral conditions, occasionally moving into a La Niña event the following fall or winter. A **La Niña** event is effectively an intensification of the normal or neutral ENSO pattern, characterized by an anomalously warm Indo-Pacific Warm Pool, enhanced upwelling in the eastern tropical Pacific, and an intensified Walker circulation. The ENSO system cycles with an

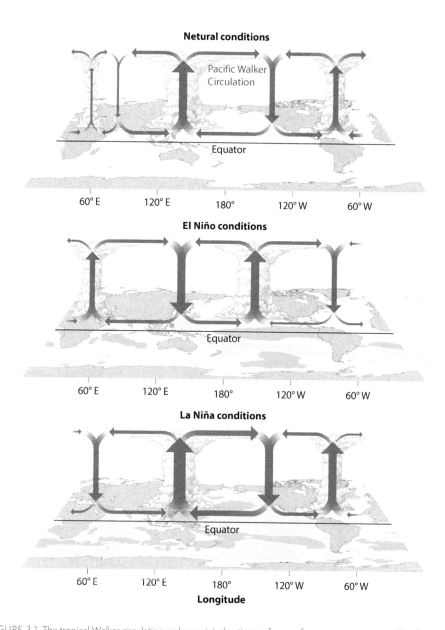

Netural conditions

Pacific Walker
Circulation

Equator

60° E 120° E 180° 120° W 60° W

El Niño conditions

Equator

60° E 120° E 180° 120° W 60° W

La Niña conditions

Equator

60° E 120° E 180° 120° W 60° W

Longitude

FIGURE 3.1 The tropical Walker circulation and associated patterns of sea surface temperature anomalies during the neutral, El Niño (warm), and La Niña (cold) phases of the El Niño Southern Oscillation (ENSO). During neutral years, heating over the Maritime Continent and Indo-Pacific Warm Pool causes strong convection, precipitation, and ascent over this region. Aloft, this air travels to the east and sinks over the cold eastern tropical Pacific before returning to the west at the surface as the trade winds. The trade winds, in turn, drive strong upwelling in the eastern tropical Pacific and push warm surface waters toward the west. These patterns in the ocean and atmosphere are intensified during La Niña events. During El Niño events, however, the Walker circulation and surface trade winds weaken: warm sea surface temperature anomalies develop in the central and eastern tropical Pacific, and the main center of convection shifts eastward away from the Maritime Continent. During both El Niño and La Niña events, changes in atmospheric circulation propagate through the global atmosphere, affecting climate and weather in many regions. *Source:* NOAA, https://www.climate.gov, drawing by Fiona Martin.

approximate periodicity of 2 to 7 years, though there have been decadal and longer periods in the historical record when the occurrence or intensity of El Niño or La Niña events has been amplified or diminished.

The influence of El Niño and La Niña is nearly global, with regions on almost every continent experiencing some shift in weather and climate during these events. For most areas influenced by ENSO, the impacts from El Niño are often (but not always) opposite in sign and magnitude from those of La Niña. El Niño events regularly suppress precipitation and cause drought (figure 3.2) over Indonesia and the Maritime Continent, northern Australia, monsoon Asia (including India and Southeast Asia), the Amazon and northern Brazil, and southern Africa. Some of the most spectacular and intense impacts can be seen over Indonesia, where El Niño–induced droughts often cause massive wildfires that blanket much of the region in smoke (figure 3.3). These fires can create

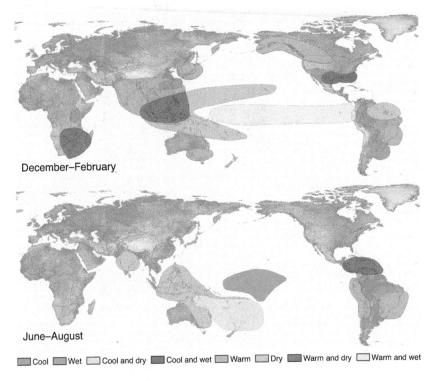

FIGURE 3.2 Typical regional climate impacts associated with El Niño, the warm phase of ENSO. El Niño events typically cause droughts over the Maritime Continent and Indonesia, northeastern Australia, southern Africa, the Amazon, and monsoon Asia. *Source:* NOAA, https://www.climate.gov.

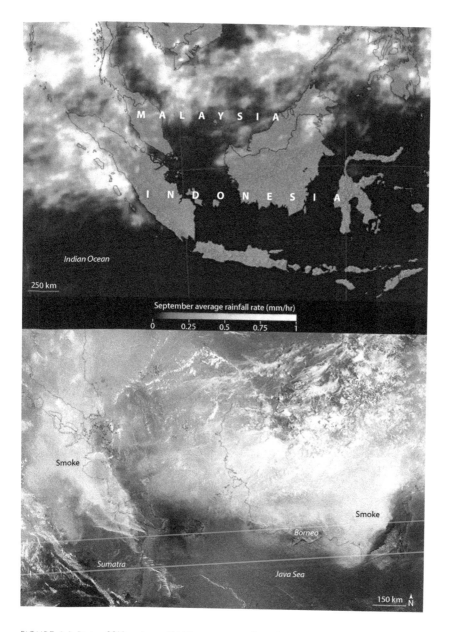

FIGURE 3.3 During 2015, a strong El Niño event caused significant drought across Indonesia and the Maritime Continent (*top panel*). This drought resulted in significant fire activity (*bottom panel*: red dots) and air pollution across the region from the resulting smoke (*bottom panel*: white areas). *Source*: NASA. Retrieved September 24, 2015, from https://www.nasa.gov/feature/goddard/2016/el-nino-brought-drought-and-fire-to-indonesia.

significant air pollution problems across the region, with significant health consequences. For example, smoke and pollution from fires in the region during an El Niño event in September and October of 2015 caused over 100,000 excess deaths in Indonesia, Malaysia, and Singapore (Koplitz et al., 2016).

The regions most likely to experience drought during La Niña (figure 3.4) are Mexico, the southwest and southern plains in North America, East Africa, extratropical South America, and the central and eastern Pacific. The southwestern United States is especially drought prone during La Niña, with many multiyear drought events in the historical record (including the 1930s Dust Bowl, the 1950s drought, and the ongoing turn-of-the-21st-century drought) being caused, in part, by a series of La Niña events. One of the most extreme recent drought years occurred in Texas and the southwestern United States during the La Niña of 2010–2011 (figure 3.5). As is typical, this

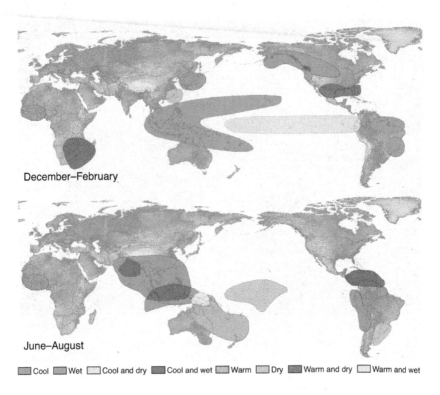

December–February

June–August

Cool Wet Cool and dry Cool and wet Warm Dry Warm and dry Warm and wet

FIGURE 3.4 Typical regional climate impacts associated with La Niña, the cool phase of ENSO. La Niña events typically cause droughts over East Africa, across Mexico, and in the southwestern, southern plains, and southeastern regions of the United States. *Source:* NOAA, https://www.climate.gov.

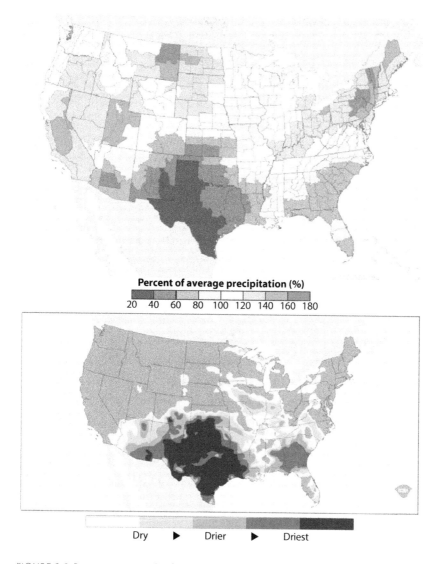

Percent of average precipitation (%)

20 40 60 80 100 120 140 160 180

Dry ▶ Drier ▶ Driest

FIGURE 3.5 Precipitation anomalies for the October 2010–September 2011 water year caused by a strong La Niña event (*top panel*). This lack of cold season moisture resulted in significant drought during the following summer of 2011 across the southwestern, southern plains, and southeastern regions of the United States, as reflected in the U.S. Drought Monitor for August 30, 2011 (*bottom panel*). *Source*: NOAA, https://www.climate.gov.

La Niña caused exceptionally low water-year (October–September) precipitation across much of the southern United States, with the most severe deficits over Texas. This lack of cold season moisture led to the development of exceptional drought conditions (the most intense category of drought in the U.S. Drought Monitor) across New Mexico, Texas, Oklahoma, Kansas, and Louisiana during the following summer. Despite its overall global importance, however, ENSO impacts in some regions of the world are relatively small or nonexistent. These regions include Europe, the Mediterranean, the Middle East, and Eurasia outside of areas affected by the monsoon.

The **Indian Ocean Dipole (IOD)** is another important pattern of ocean variability, causing drought and hydroclimate shifts in land areas surrounding the Indian Ocean (Saji et al., 1999). The IOD is a seesaw in ocean surface temperatures along the equator in the Indian Ocean, with changes in this temperature gradient affecting circulation in the atmosphere across the basin. During positive IOD events, ocean temperatures off the coast of East Africa are anomalously warm, and ocean temperatures in the east are relatively cool. This causes enhanced convection over East Africa and suppresses precipitation over the Maritime Continent, causing drought over the Maritime Continent and Australia and increased precipitation (and often flooding) over East Africa. Anomalies in the ocean and atmosphere are reversed during negative IOD events, leading to drought conditions over East Africa and India. Importantly, both the IOD and ENSO share a common center of action near the Maritime Continent. When the IOD and ENSO are in phase (positive IOD with an El Niño), subsidence over this region is further enhanced, and precipitation deficits and drought impacts are amplified.

Other modes of ocean variability operate on longer timescales, with positive or negative phases of these modes persisting for often a decade or longer. Similar in spatial extent and impact to ENSO are decadal variations in Pacific Ocean temperatures, often referred to as the **Interdecadal Pacific Oscillation (IPO)** or **Pacific Decadal Oscillation (PDO)** (Dong & Dai, 2015; Mantua & Hare, 2002). The spatial patterns of ocean temperature anomalies during positive IPO/PDO events closely resemble El Niño events, and negative IPO/PDO events are similar to La Niña events. Unlike ENSO, however, the IPO/PDO changes slowly from year to year, with any given phase typically persisting for one to two decades. For example, the IPO/PDO was locked in a sustained negative phase from the 1950s to the late 1970s before switching to a positive phase from ~1980 to 2000. Because of this persistence, IPO/PDO impacts on drought and hydroclimate may be similarly sustained. For example, the probability of drought occurrence in the southwestern United States is substantially higher if the IPO/PDO is in a negative phase (McCabe et al., 2004). Because of the similarity in spatial structure between ENSO and the IPO/PDO, the global impacts of the IPO/PDO on hydroclimate are broadly similar to those of ENSO, with amplified effects when ENSO and IPO/PDO events are in phase (positive IPO/PDO with El Niño and negative IPO/PDO with La Niña).

Operating on even longer timescales is another pattern of ocean variability, the **Atlantic Multidecadal Oscillation (AMO)**. The AMO is related to circulation changes in the North Atlantic and exhibits even greater persistence than the IPO/PDO. The AMO was in an extended warm, or positive, phase from 1930 to 1960 and a cool, or negative, phase from 1970 to the late 1990s. During warm phases of the AMO, drought is favored across much of North America, especially over the Great Plains and Mississippi River valley. A warm tropical Atlantic, associated with the positive AMO, weakens the transport of moisture into the continent from the Gulf of Mexico, especially during the summer and fall seasons. Warm ocean temperatures associated with positive phases of the AMO also shift the Intertropical Convergence Zone northward, increasing precipitation across West Africa and inducing anomalous subsidence and drought across northern tropical South America.

Although these ocean-atmosphere teleconnections can offer significant predictive power, the regional hydroclimate response is not deterministic. This is because of the nonstationary nature of many of these teleconnections, a consequence of the myriad other ongoing processes in the climate system that also affect regional hydroclimate. Such instabilities and breakdowns in the expected strength or sign of the teleconnections may occur because of internal variability in the atmosphere unrelated to the oceans, local land-atmosphere interactions, and even forcing from other ocean basins. For example, precipitation in Mexico is sensitive to ENSO, but sea surface temperature variability in the tropical Atlantic also has a strong regional impact (Seager et al., 2009; Stahle et al., 2016), so the regional hydroclimate response will depend on forcing from both basins. Similarly, the impact of ENSO events in Australia may be stronger or weaker depending on the concurrent phasing of the IOD and IPO (W. Cai et al., 2010; Power et al., 1999; Risbey et al., 2009). Impacts of these teleconnections, as powerful as they are, should therefore be viewed primarily in a probabilistic sense. A La Niña event may make it much more likely for southwestern North America to experience a drought, but is by no means a guarantee.

The Land Surface

Land surface processes and interactions with the atmosphere play important roles in drought dynamics. To a first order, the soil acts as an integrator of moisture supply (precipitation) and demand (evapotranspiration), imparting persistence or "memory" into the climate system because it can carry moisture deficits or surpluses forward in time from one season to the next or even from year to year. This is illustrated clearly through the sensitivity of tree growth (represented by growth patterns in tree rings) to precipitation during the winter and summer (figure 3.6) (St. George & Ault, 2014). Each triangle represents the relationship (correlation) between the growth of trees (which occurs during the boreal spring and summer) and precipitation from the previous winter or concurrent summer. For much of the Northern Hemisphere, growth is most strongly related to precipitation during the summer. Over much of the western United States (especially California and the Southwest), however, growth is much more strongly correlated with precipitation from the previous winter than with that during the current growing season. In this region, winter temperatures are cooler, and evaporative demand is substantially lower compared to the growing season. Precipitation during this season thus accumulates in the soil (or snowpack) and is then carried over to be used by trees during the following summer. Precipitation during the summer is less important because temperature and evapotranspiration rates are higher. As a result, less of this precipitation is effectively available to recharge the soils and contribute to vegetation and tree growth.

Interactions between the surface and the atmosphere can also intensify droughts and cause additional impacts, such as heat waves, because surface moisture availability strongly controls exchanges of energy and water between the surface and the atmosphere. When temperatures are warm and evaporative

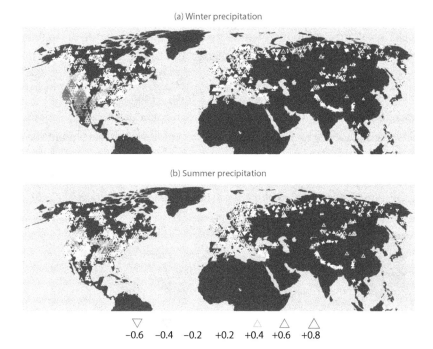

(a) Winter precipitation

(b) Summer precipitation

▽   △   △   △
−0.6   −0.4   −0.2   +0.2   +0.4   +0.6   +0.8

FIGURE 3.6 Correlations between local records of tree growth (tree-ring chronologies from the International Tree-Ring Data Bank) and precipitation from the previous winter and concurrent summer. Over most of the Northern Hemisphere, the warm season (April–September) is the primary growing season. Despite this, in many regions (e.g., southwestern North America), tree growth is most sensitive to winter season precipitation. This asynchrony reflects the fact that this cold season precipitation is stored as soil moisture and carried over into the growing season, when it is used by the trees. *Source*: Scott St. George, adapted from Figure 1, St. George & Ault, 2014.

demand is high, evapotranspiration rates are strongly limited by moisture supply at the surface. As the soils begin to dry out, actual evapotranspiration will be unable to satisfy the atmospheric demand, and the energy balance will shift to favor sensible heating (the energy fluxes associated with warming) over latent heating (the implicit energy associated with the phase change of water). This translates to a reduced moisture flux from the surface to the atmosphere and increased heating of the soil and air. Soil moisture deficits associated with droughts can therefore play a critical role in amplifying heat extremes and heat waves, with warmer temperatures in the atmosphere further increasing evaporative demand and moisture losses from the surface.

This connection between droughts and heat waves is illustrated in figure 3.7, which shows correlations between the number of hot days during the hottest month of the year and precipitation anomalies over the preceding 3-, 6-, and 9-month periods. Over many regions (e.g., the United States, Mexico, western and eastern Europe, northeast China, South America, southern Africa, and Australia), there are strong and significant negative correlations, indicating a tendency for hot conditions to occur following several months of below-normal precipitation. This mechanism was also clearly demonstrated during the European drought and heat wave of 2003, where extreme heat during the summer caused ~40,000 excess deaths (García-Herrera et al., 2010).

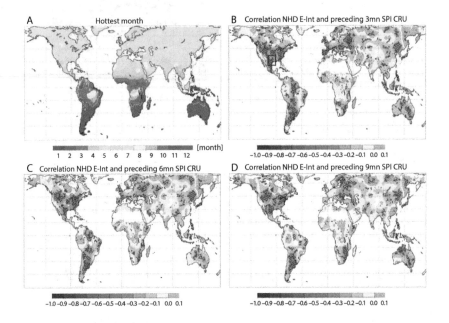

FIGURE 3.7 The hottest month per year (January through December) for each location (panel A) and the correlations between the number of hot days in this month (days with a maximum temperature exceeding the 90th percentile) and precipitation for the preceding 3 months (panel B), 6 months (panel C), and 9 months (panel D), represented by the Standardized Precipitation Index. Hatched regions are areas where the correlation is significant (90 percent confidence level), which include large areas of North America, Europe, and South America. Thus, for many regions, anomalously hot summers often occur following meteorological drought conditions in the previous seasons. Source: Reproduced, with permission, from Figure 1, Mueller & Seneviratne, 2012.

The 2003 European drought was caused initially by low spring precipitation, which carried over into the summer as a significant soil moisture drought. With little water to evaporate, the energy balance shifted overwhelmingly toward sensible heating, driving the strong heat wave that persisted through most of July over western Europe. Indeed, computer model simulations are able to reproduce many important characteristics of this event only when land-atmosphere interactions and the influence of the soil moisture drought are included (Fischer et al., 2007).

Soil moisture can also influence drought by directly affecting precipitation through two possible mechanisms. The first is *precipitation recycling* (Eltahir & Bras, 1996), where local to regional evapotranspiration from the land surface provides a significant fraction of the total atmospheric moisture budget. In cases of drought where soil moisture and evapotranspiration from the land surface are significantly reduced, subsequent precipitation may be further reduced because less overall moisture is available. More recent research, however, has focused less on this mechanism of precipitation recycling and more on how soil moisture affects processes in the atmosphere that trigger precipitation, regardless of the moisture source. For example, high evapotranspiration rates associated with wet soils can help destabilize the atmosphere, increasing the likelihood of convection and precipitation even if most of the atmospheric moisture providing the precipitation originates from more distant regions, such as the oceans. There are, however, still large uncertainties regarding the strength (and even the sign) of soil moisture–precipitation interactions for many areas. This is due, in part, to the relative paucity of long-term soil moisture observations and the difficulty in separating the relatively smaller effect of soil moisture on precipitation from the much larger direct influence of precipitation on soil moisture (Seneviratne et al., 2010). Despite these uncertainties, several regions are considered hot spots of land-atmosphere interaction: areas where the soil moisture state has a strong influence on precipitation. These include regions that were initially highlighted by the seminal study of Koster et al. (2004), including the central plains of North America, West Africa, and India.

An important modulator of land-atmosphere interactions and feedbacks, especially from a drought perspective, is the terrestrial vegetation. To a first order, the global distribution of different vegetation and ecosystem types (e.g., grasslands, rainforests, deciduous broadleaf forests, and tundra) is controlled by climate, especially temperature and precipitation. Like soil moisture, however, the state of the vegetation can have significant impacts on local weather and climate by modulating interactions between the land surface and atmosphere (Bonan, 2008). The amount and type of vegetation affect nearly all aspects of the surface energy balance (net longwave and shortwave fluxes and sensible and latent heating) and moisture budget (e.g., evapotranspiration, interception, infiltration, and runoff). Vegetation has a typically lower albedo than bare soil, and forests have lower albedos than grasslands, especially in regions and seasons with significant snow cover where the taller trees mask higher-albedo snow below the canopy. A lower albedo typically means greater absorption of energy at the surface, which can make increased energy available for evapotranspiration, sensible heating, and convection. Through transpiration and interception by the canopy, vegetation is also the primary conduit for latent heating, which supplies moisture to the atmosphere and (ultimately) precipitation. Trees and forests, with typically greater leaf areas and deeper roots than grasslands or shrublands, have higher rates of evapotranspiration compared to other vegetation types. Vegetation also adds roughness to the surface, which can block winds, absorb momentum from the atmosphere, and generate mechanical turbulence, forcing air to mix and rise and contributing to cloud formation and precipitation. Losses of vegetation through deforestation, land degradation, or other processes can thus potentially influence regional climate and precipitation patterns through a variety of mechanisms.

One of the first scientists to propose an important role for vegetation in drought dynamics was Jule Charney, author of a 1975 study investigating a drought of nearly unprecedented severity that occurred in the Sahel region of West Africa in the 1970s (this drought is covered in more detail in chapter 6). Using a climate model (a computer program designed to simulate processes in the climate system), he simulated a reduction in vegetation coverage over the region by increasing surface albedo from 14 percent to 35 percent. This albedo increase caused a weakening of the West African monsoon in the model, which reduced precipitation over the Sahel by 40 percent, and Charney concluded that **land degradation/desertification** in the region was a major contributor to this drought. Later work would show that Charney's analysis of the actual impact of this vegetation-albedo mechanism was oversimplified and that the role of vegetation dynamics in the Sahel drought was likely secondary to ocean temperature changes at the time. Still, this study helped establish the importance of considering land surface and vegetation processes in climate and drought research.

Vegetation-atmosphere interactions are also believed to have played a major role during the mid-Holocene (~9,000 to 6,000 years before present), when the West African monsoon penetrated deep into the Sahara Desert, leading to wetting and the expansion of savanna ecosystems across the region (the so-called **Green Sahara** period). This wetting was likely forced, initially, by increased summer solar energy inputs at the time that caused strong heating over northern Africa, enhancing the monsoon. It is also hypothesized, however, that the expansion of vegetation with the monsoon provided a strong positive feedback (Claussen et al., 1999), reducing the surface albedo and increasing precipitation recycling, thus helping to maintain the strength and expansion of the monsoon for thousands of years. Later, as solar energy inputs declined, the monsoon began to weaken and retreat to the south, causing the vegetation to die and the monsoon to weaken even further; eventually, the region experienced an abrupt collapse toward the extreme desert conditions that are more typical today.

Finally, vegetation can interact with other important components of the Earth system, such as mineral dust, with potentially important ramifications for drought dynamics. When soils are eroded from the surface by the wind, some of the dust particles can remain suspended in the atmosphere for days as aerosols. In the modern period, the largest fluxes of these dust aerosols come from dry desert regions with little or no vegetation, such as the Bodélé Depression in North Africa (Engelstaedter et al., 2006). However, losses of vegetation in regions with strong winds and a propensity for drought (as dry soils are more easily eroded) can also generate significant dust emissions until the vegetation recovers. Potential feedbacks to drought occur because dust aerosols are highly reflective in the visible spectrum, increasing overall albedo, reflecting more solar energy back to space, and reducing overall energy availability for evapotranspiration, convection, and precipitation. This interaction between dust and drought has been proposed as one hypothesis to explain the almost unprecedented nature of the Dust Bowl drought over the United States in the 1930s (B. Cook et al., 2009). The decades leading up to the 1930s saw a significant expansion of U.S. agriculture into the central plains, leading to the replacement of the more deeply rooted, perennial, and drought-resistant prairie grasses with drought-vulnerable wheat. When a drought began in the 1930s, likely forced initially by a La Niña event, the crops failed spectacularly, causing widespread losses of vegetation cover, massive wind erosion, and high levels of dust aerosol concentrations across the region. This dust aerosol forcing, along with reductions in surface vegetation, acted to amplify the drought by increasing temperatures and further suppressing precipitation, turning this event into one of the worst natural disasters in the history of the country. Similarly, interactions between vegetation and dust aerosols likely played an important role in the collapse of the Green Sahara (Tierney et al., 2017).

The Atmosphere

Circulation patterns in the atmosphere that cause drought do not always arise from changes in the oceans or interactions with the land surface. The atmosphere itself has its own intrinsic internal variability that can result in the development of conditions that favor drought (i.e., high pressure and ridging) independent of any exogenous forcing. Notably, the typical spatial and temporal scales of internal atmospheric variability are often markedly different from patterns forced by the ocean or land surface. Forcing from ocean modes can span hemispheres or continental-scale regions, resulting in coherent drought anomalies across large areas. ENSO events, for example, cause a widespread reorganization of circulation across the tropics and mid-latitudes, often producing simultaneous droughts in geographically distant regions. By contrast, atmospheric variability tends to be much more localized: a strong, internally generated ridge over western North America, for example, is unlikely to stimulate responses over distant regions like Indonesia or East Africa. The atmosphere itself also has very little intrinsic memory or persistence compared to the ocean and land surfaces. Internally generated atmospheric circulation anomalies typically persist from days to (occasionally) several months. Maintenance of such circulation anomalies and the associated drought conditions beyond this timeframe (e.g., into the next season or year) usually requires some contribution from the ocean or land surface. Internal atmospheric variability affects drought dynamics every-where, but it is typically more important over regions where teleconnections to ocean forcing and land-atmosphere feedbacks are weaker. In the Northern Hemisphere, this includes central and eastern Canada, northern Eurasia, and central Europe (Schubert et al., 2016).

One example of a severe drought forced primarily by internal atmospheric variability is the most recent drought in California. Over this region, the cold season (October–March) is the main season of moisture supply. For much of the drought, however, circulation during these months was characterized by a large and persistent ridging pattern in the atmosphere centered along the Pacific Coast, especially during the worst years of the drought from 2012 to 2015 (figure 3.8). Dubbed the "Ridiculously Resilient Ridge (RRR)" for its

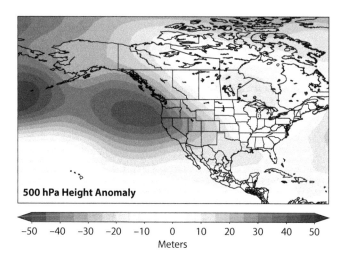

500 hPa Height Anomaly

−50   −40   −30   −20   −10   0   10   20   30   40   50
Meters

FIGURE 3.8 Geopotential height anomalies (relative to a 1981–2000 baseline) at 500 hectopascals (hPa) for October–March during the worst years of the most recent California drought (2012–2015). These winters were characterized by persistent ridging and high-pressure conditions centered along the west coast of North America. This circulation feature, arising primarily from internal atmospheric variability, deflected storms and suppressed precipitation across California. *Source:* Data from NCEP-NCAR Reanalysis 1, provided by the NOAA-ESRL Physical Sciences Division, Boulder, Colorado, https://www.esrl.noaa.gov/psd.

exceptional persistence (Swain et al., 2014), this feature suppressed precipitation and deflected storms away from the coast, including California, resulting in a succession of years of steadily intensifying precipitation deficits and drought. Internal atmospheric variability has also been diagnosed as a major contributor to other recent drought events, including the 2010 drought and heat wave over Moscow and western Russia (Schubert et al., 2014) and the 2012 summer drought over the central plains of the United States (Hoerling et al., 2014). Even in cases where internal atmospheric variability dominates, however, other factors may still play some role. For example, there is some evidence that remote forcing from tropical Pacific sea surface temperatures may have contributed to the anomalous persistence of the RRR, even though it arose primarily through internal atmospheric variability (Seager et al., 2015; Swain et al., 2017).

As with the oceans, there are several preferred and recurrent modes of variability in the atmosphere relevant for drought dynamics. Of these, possibly the most well studied and best understood is the **North Atlantic Oscillation**, or NAO (Hurrell et al., 2001). The NAO pattern is centered in the North Atlantic and is most active in boreal winter and spring. Its strength and polarity are modulated by the pressure gradient between the semipermanent Icelandic Low at high latitudes and the Azores High in the subtropics. During a positive NAO, the pressure gradient between the Azores High and Icelandic Low is strong, deflecting winter storms into northern Europe and causing relatively warm and wet conditions in that region and drought in the Mediterranean. During a negative NAO, this pressure gradient is weaker, and the preferred track for storms is into the Mediterranean, resulting in dry conditions in northern Europe. Patterns that are similar to the NAO but that affect much larger areas across the Northern Hemisphere include the Northern Annular Mode and Arctic Oscillation (Thompson & Wallace, 2001), both of which have corollaries in the Southern Hemisphere (the Southern Annular Mode and Antarctic Oscillation). As with most patterns of internal atmospheric variability, persistence in the NAO (and related patterns) is relatively weak, and pattern shifts are generally predictable only several days, and sometimes weeks, in advance. Other important modes of atmospheric variability that have important impacts for hydroclimate include the Pacific/North American pattern (affecting western North America) (Leathers et al., 1991) and the Madden-Julian Oscillation (primarily influencing the tropics and monsoon regions) (Zhang, 2005).

Flash Droughts

Flash droughts are drought events (usually defined in terms of soil moisture) that develop rapidly on timescales of day to weeks, typically more quickly than conventional droughts (Otkin et al., 2017). They often occur with little to no advance warning, making it difficult to anticipate and prepare for them, and, as a result, they can cause severe agricultural losses. Like most droughts, flash droughts are usually forced by precipitation deficits, but evaporative losses associated with heat waves can enhance and contribute to the development of these events. Flash droughts are most common during the growing season, when background levels of evaporative demand are already high and at least average levels of precipitation are needed to maintain soil moisture.

A canonical example of an extreme flash drought occurred during the summer of 2012 over the central plains of the United States and can be seen in a variety of drought indicators over the region (Otkin et al., 2016). In May, mostly normal levels of precipitation over much of the region resulted in only localized drought conditions in some areas and overall above-average yields for winter wheat. Beginning in June, however, lack of precipitation resulted in a rapid expansion and intensification of drought conditions, causing significant declines in soil moisture and increases in evaporative stress. Because the rapid drought intensification occurred after the main summer planting of corn and soybeans, agriculture experienced significant negative impacts. Corn yields were 26 percent below forecasted expectations issued by the U.S. Department of Agriculture at the beginning of the growing season, and economic losses by the end of July were $12 billion (Hoerling et al., 2014). The drought largely abated by the following growing season, but 2012 highlighted the damage that intense and rapidly developing flash droughts could inflict.

Snow Droughts

Snow is an integral part of seasonal hydrology in many mountainous regions, with runoff pulses associated with spring snowmelt providing critical water resources to populations and ecosystems occupying lower elevations. Included in the United States are the Cascade Range in the Pacific Northwest, the Sierra Nevada in California, and the Rocky Mountains in Colorado and the Interior West. A **snow drought** occurs when the snowpack is below average (Harpold et al., 2017). This can occur when there are deficits in seasonal precipitation but also when precipitation is normal but temperatures are anomalously warm. In the latter case, often referred to as a "wet" snow drought, warmer temperatures increase the fraction of precipitation falling as rain and increase snowpack losses during the cold season from melting and sublimation.

Wet snow droughts typically translate to less total available soil moisture and lower overall streamflow during the growing season, even if the total seasonal precipitation is near normal. In addition, runoff during wet snow droughts is typically more evenly distributed across the cold season, when it is largely unavailable to dormant vegetation. In cases where reservoirs are operated to manage for flooding during the winter and capture the seasonal snowmelt in the spring, this can mean declines in reservoir storage moving into the warm season. Most recently, a major wet snow drought affected the Pacific Northwest in 2015. From October 2014 to March 2015, total precipitation across Oregon and Washington was near normal, but temperatures climbed to record highs. As a consequence, the snowpack reached its lowest level across much of the region (Mote et al., 2016), leading to intense drought during the spring and summer of 2015, as recorded in the U.S. Drought Monitor.

Predictability of Droughts

Predicting climate events, including droughts, requires exploiting aspects of the climate system that have significant persistence or memory, such that current conditions provide useful information to constrain the state of the system in the future. For monthly to seasonal forecasts, the typical persistence of circulation anomalies in the atmosphere alone is too short, usually providing little predictive capability beyond about 10 days (the maximum useful length of most weather forecasts). Other components of the climate system, such as the oceans and the land surface, have much longer memory and much greater potential to provide useful information for predictions several months to seasons in advance.

To date, the greatest advances in seasonal climate prediction use ocean temperatures, especially those associated with ENSO (for example, see the seasonal forecasts provided by the International Research Institute for Climate and Society, https://iri.columbia.edu/our-expertise/climate /forecasts/seasonal-climate-forecasts/). El Niño and La Niña events typically peak in strength in November or December, but their development can be tracked in ocean temperatures in the tropical Pacific many months ahead of time. Using this information in forecast models, ENSO events and their expected regional impacts on precipitation and droughts can then be predicted months in advance. The recent Indonesian drought in the summer and fall of 2015, for example, was anticipated well ahead of time because warming ocean temperatures in the eastern tropical Pacific earlier in the year indicated that an El Niño event was developing (e.g., Shawki et al., 2017).

Other modes of ocean variability, such as the IPO/PDO and AMO, may provide additional predictability on even longer timescales. For example, if a persistent negative phase of the PDO occurs over the next decade, the average climate can be expected to be drier than normal (and droughts more likely) in regions like southwestern North America. Exploiting other modes beyond ENSO for predictability purposes, however, is still difficult because our understanding of the underlying physics and dynamics of these modes is much more uncertain. Soil moisture conditions may also provide some significant predictive skill, either through the simple carryover of soil moisture anomalies from season to season or through land-atmosphere interactions that influence precipitation or evapotranspiration. The importance of spring soil moisture, for example, has been demonstrated after the fact for the 1988 and 2012 droughts and the 1993 floods over the central plains of the United States (Saini et al., 2016). The development of consistently skillful seasonal forecasts using soil moisture information is still largely a work in progress, however, because of the uncertainties in the underlying land surface processes (including their representation in models) and the paucity of high-resolution and high-quality soil moisture observations.

Ecosystem Impacts

The consequences of drought for ecosystems depend on a variety of factors. In regions where drought is a regular or cyclic occurrence, local ecosystems may be highly adapted to recurring shifts in moisture availability, making them less susceptible to droughts. However, when the most extreme conditions occur (e.g., a drought that persists for a decade or more, a drought event combined with intense heat), or where drought is a rare occurrence, the adaptive capacity of species and ecosystems may be overcome, resulting in significant morbidity, mortality, and even shifts in ecosystem structure and function. The full impact of a given drought event therefore depends on the duration and severity of the associated moisture deficits and the resilience of the ecosystems themselves.

Drought impacts on ecosystems typically cascade from leaf-level processes (e.g., photosynthesis and stomatal conductance) to the whole plant and ecosystem (van der Molen et al., 2011). In response to drought, most ecosystems experience a decline in both productivity (carbon gain by the plants via photosynthesis) and respiration (carbon loss from plant respiration and heterotrophic decomposition). Productivity losses occur because many plants close their stomata in response to moisture deficits, reducing stomatal conductance to limit water losses. The stomatal closure, however, simultaneously reduces carbon diffusion into the leaf, limiting photosynthesis and productivity. The degree of stomatal closure in response to drought can vary widely from species to species. Isohydric species will strongly limit stomatal conductance in response to drought, experiencing more severe short-term reductions in productivity but also increased water savings compared to anisohydric species, which have very little control over their stomata. Vegetation may also undergo structural changes in response to drought, including senescence of leaves, that may further reduce productivity. Respiration declines during drought have been linked primarily to the reduced availability of carbon compounds as a consequence of declines in both productivity and microbial activity in soils. During droughts, productivity declines generally exceed the reductions in respiration; most ecosystems therefore become net sources of carbon to the atmosphere during drought events.

Productivity losses during droughts can carry forward in time as reduced carbohydrate reserves in plants. This often translates to reduced plant growth in the years during and following a drought (Anderegg et al., 2015), as well as increased vulnerability to pathogens (e.g., diseases, fungi, and insects) and other disturbances (e.g., fire and wind). Vegetation mortality from pathogens or disturbances is an indirect consequence of drought, but drought is hypothesized to induce vegetation death directly through two main processes. The first is **hydraulic failure** from cavitation of the xylem vessels, which carry water within the plant. This occurs when canopy moisture demands exceed the available supply from the stem and soil, resulting in an aspiration of the water column in the xylem. Direct mortality has also been hypothesized to occur through **carbon starvation**. This occurs when productivity and carbohydrate reserves drop to the level where basic metabolism can no longer be maintained. A recent analysis of existing literature suggests that, for trees, hydraulic failure is a much more universal and widespread mechanism for drought mortality than carbon starvation (Adams et al., 2017).

Vegetation mortality events associated with droughts can be quite large and have far-reaching consequences. In the 1950s, a decadal drought event (caused by a series of La Niña events) affected much of the southwestern United States. Over New Mexico, this drought caused a widespread die-off of ponderosa pine forest, which was replaced by piñon pine and juniper woodland (Allen & Breshears, 1998). Notably, this ecosystem shift has persisted for decades, even after the drought ended and moisture levels returned to normal. In the western United States, persistent drought conditions from 2000 to 2013 contributed to a significant increase in mountain pine beetle populations, which then caused tree deaths over 71,000 km² of pine forest (Hart et al., 2015). More recently, the U.S. Forest Service estimated that by the fall of 2016, the 2012–2016 drought in California had killed over 100 million trees in the state (Stevens, 2016). In a rare example of a case with global impacts, a series of concurrent droughts in the Southern Hemisphere in the early 2000s significantly reduced plant productivity across broad regions of South America, southern Africa, and Australia (Zhao & Running, 2010). This anomaly was significant enough to counteract increased vegetation growth in the Northern Hemisphere, resulting in a temporary increase in atmospheric carbon dioxide growth rates (Zhao & Running, 2010).

FOUR

Drought and Hydroclimate in the Holocene

aleoclimatology is the study of past climate variability and change, focused primarily on time intervals prior to the last 150 years of widely available instrumental measurements. In this chapter, we will discuss drought and hydroclimate during the **Holocene**, the current interglacial period, which began ~11,700 years ago and during which some of the most significant technological developments (e.g., agriculture) and social developments (e.g., the shift from mobile hunter-gatherer lifestyles to sedentary agricultural societies) in the history of humanity occurred. This time interval therefore provides a useful baseline for defining natural climate variability to properly contextualize the direction, magnitude, and impacts of anthropogenic climate change. Holocene paleoclimate, however, encompasses an impressively broad and deep array of datasets and phenomena. Given the impossibility of covering everything in detail, we will therefore focus on major hydroclimate events and attempt to provide answers to some of the most fundamental questions related to hydroclimate in the paleoclimate record. How has regional hydroclimate changed over the last 11,700 years? How do droughts in the paleoclimate record compare to more contemporary events? What impact did past droughts have on people and societies? And how can we use paleoclimatology to inform our understanding of water resources in the modern era?

Why Paleoclimate?

Paleoclimatology is a field full of compelling research questions, but its relevance for investigating modern and future climate change may not be immediately apparent. Broadly speaking, paleoclimatology provides additional information on climate variability and change beyond the relatively short instrumental record, facilitating a much more comprehensive view of the climate system. This perspective is especially important for understanding extreme events like droughts, which are (by their very definition) relatively rare occurrences that may be underrepresented in shorter instrumental datasets of the last 150 years. Among other applications, paleoclimate data can improve our estimates of the range of natural variability in the climate system, show how this variability has changed over time, better constrain the risks and return periods of extreme events, and place recent observed changes in a longer-term context.

For example, the severity of the 2012–2016 drought in California has led to speculation that climate change may be amplifying the severity of drought in the region (Diffenbaugh et al., 2015). One important indicator of drought and water variability in California is the Sierra Nevada snowpack on April 1, as it provides the meltwater that refills most of the major reservoirs in the state for use during the summer dry season. One indicator of the snowpack, snow water equivalent, was recently reconstructed for the last 500 years using tree rings (figure 4.1). This reconstruction clearly shows that, in 2015, April 1 snow water equivalent in the Sierra Nevada was the lowest of the last 500 years, the most severe snow drought in the record. Tree-ring-based reconstructions of soil moisture similarly suggest that this most recent drought in California may have been the most intense of the last millennium (Griffin & Anchukaitis, 2014). Although not conclusive on their own, these studies support the idea that climate change may already be affecting droughts and starting to push the system outside of the range of natural variability.

Another clear situation where information from the paleoclimate record is valuable for understanding modern climate variability is that of the Colorado River Basin, one of the most intensely managed basins in terms of water resources in the world. In 1922, the Colorado River Compact apportioned consumptive water use among all states in the river basin: Wyoming, Utah, Colorado,

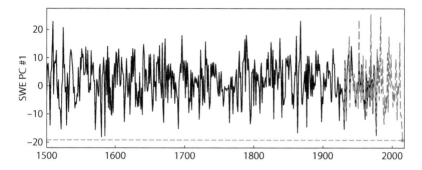

FIGURE 4.1 Reconstructed (black line) and observed (red line) snow water equivalent (SWE) for April 1 in the Sierra Nevada from 1500 to 2015. This reconstruction represents the leading mode of variability in April 1 SWE over the Sierra Nevada, which accounts for 63 percent of the total underlying variance. The reconstruction shows that, by this indicator, April 1 SWE in 2015 was likely the lowest of the last 500 years. *Source:* Data from Soumaya Belmecheri; figure adapted and modified from Figure 1, Belmecheri et al., 2016.

New Mexico, California, Arizona, and Nevada (Christensen et al., 2004). A later agreement in 1944 further guaranteed a certain level of flow to Mexico every year. The water allocations in these agreements were based on instrumental flow measurements of the Colorado River at Lee's Ferry from the early part of the 20th century, when the average flow was estimated at 22 billion cubic meters (BCM) per year. This, however, turned out to be an overly optimistic baseline because this period was exceptionally wet over much of the western United States, including the Colorado River Basin (B. Cook et al., 2011). Measurements over a much longer instrumental period (1906–2000) show that long-term average flow was lower, only 18.6 BCM per year, with substantial variability from year to year (Christensen et al., 2004). Paleoclimate reconstructions of flow from tree rings further demonstrate that the early 20th-century flows used as a basis for the Colorado River Compact were likely the highest for any time in the last 500 years, when long-term average flows (16.7 BCM) were even lower than the 20th-century mean (Meko et al., 2007). The lack of a longer-term perspective provided by the paleoclimate record allowed an unrealistic view of the true availability of water in the basin, leading to overly generous water allocations relative to what the climate system could support.

Ultimately, paleoclimatology represents an additional set of tools we can use to improve our understanding of how the climate system works and what changes we can expect in the future. Paleoclimate cannot answer all questions, of course, and there are various levels of uncertainties that must be considered in interpreting the proxy records. But used in conjunction with models, observations, and theory, paleoclimatology can provide an important long-term context and test bed for our understanding of the climate system.

Paleoclimate and People

One major area of research to which paleoclimatologists have contributed involves the interactions between past human societies and the climate system. This is especially true for hydroclimate, where many researchers have connected major societal shifts or even collapses (e.g., the Maya in Central America and the Angkor in Southeast Asia) with significant climate events, especially droughts. But even though it may be tempting to draw direct connections between major climate events and changes inferred from the often-fragmentary archaeological and paleoclimate record, the causal links are rarely straightforward. Even when implicated in major social upheavals, environmental events are unlikely to be the sole causes of societal collapses. Rather, they most likely act as additional stressors on social systems that may already be experiencing strain from internal strife, reduced availability of food and other resources, or conflicts with other societies. The term *collapse* is also imprecisely defined and has been used to describe a range of phenomena, from the abandonment of local settlements to the complete dissolution of regional polities. Throughout this chapter, we will highlight cases where there is at least some evidence that hydroclimate changes in the paleoclimate record have impacted societies and people. In certain cases, however, there may be substantial expert disagreement on the causes, linkages, and level of interaction between environmental events and the associated social changes.

Paleoclimate Proxies

Information on the climate system prior to the availability of instrumental measurements (e.g., thermometers and rain gauges) is derived primarily from *climate proxy* records. These are natural (primarily biological and geo-logical) archives that can be used to infer changes in temperature, moisture, atmospheric composition, and other climate system variables. Among the many examples are ice cores, tree rings, lake and ocean sediments, speleo-thems (secondary mineral deposits in caves), and corals. Different proxies do not all record the same information, and no single proxy is equally suitable for all applications. For Holocene hydroclimate, the most useful proxies are those that (1) resolve time intervals of 1,000 years or less and (2) respond to or record changes in the hydrologic cycle directly (e.g., through precipitation or runoff) or indirectly (e.g., through changes in vegetation distributions or productivity).

Studies of Holocene paleoclimate can somewhat arbitrarily be divided into two categories. The first is concerned with climate evolution over the entire Holocene, from ~11,700 years ago to today. At this scale, global cli-mate changes are dominated by the receding ice sheets from the last glacial period and orbitally driven shifts in the seasonal and latitudinal distributions of insolation. Proxies of sufficient length to cover significant stretches of the entire Holocene include ice cores, accumulated sediments in lakes and coastal regions, and speleothems. In some cases, proxy records may even extend across the entire Holocene or beyond, providing continuous singular measurements of climate up to the present day. The extreme length of these records, however, often comes at the cost of time resolution, with many such proxies unable to distinguish time periods at a finer resolution than one to several decades.

The second category of studies is focused on the last 2,000 years, referred to as the Common Era (CE). Dominant natural forcings over these centu-ries include variability in solar output, volcanoes, land-use and land-cover change, and (especially over the last 150 years) greenhouse gases (GHGs). For these centuries, many proxies are available that can resolve individual years (e.g., tree rings), allowing for more precise dating of major climate events in the past. However, because many of these more precise proxies cover relatively short time intervals (e.g., trees rarely live more than several hundred or a thousand years), their use for studying Holocene hydroclimate is largely limited to the most recent centuries.

Orbital Forcing and Climate Evolution from the Last Ice Age

The Holocene is our current geological epoch, the latest warm interval in the recurring cycles of colder stadial (glacial) and warmer interstadial (interglacial) periods that began with the onset of the Pleistocene ~2.6 to 2.7 million years ago. The most recent stadial peaked at the Last Glacial Maximum (LGM) ~20,000 to ~25,000 years ago, when extensive ice sheets covered much of northern North America and Eurasia. The Holocene itself began ~11,700 years ago, following the Younger Dryas, a temporary period of abrupt cooling in the Northern Hemisphere that interrupted the long-term warming trend out of the LGM. Since then, the long-term climate evolution of the Holocene on millennial timescales has been driven by changes in the extent of ice sheets and orbital forcing (Wanner et al., 2008, 2015).

Orbital forcing (changes in the orbital configuration of the Earth-Sun system induced by the gravitational pull of various celestial bodies, including Jupiter and Saturn) was the initial cause of the warming that brought the Earth out of the last ice age. The components of orbital forcing most relevant from a climate perspective are eccentricity (changes in the ellipticity of Earth's orbit; periodicity of 100,000 years), obliquity (changes in the angle of the Earth's tilt relative to the plane of orbit; periodicity of ~41,000 years), and axial precession (changes in the direction of the axis of Earth's rotation relative to the fixed stars; periodicity of ~26,000 years). Orbital forcing has little effect on the integrated global or annual solar energy budget on Earth but can significantly change the latitudinal and seasonal distribution of insolation. During the transition from the LGM to the Holocene, orbital changes (occurring primarily through the interaction between the obliquity and precession cycles) increased boreal summer (July) insolation in the Northern Hemisphere. The increased insolation and the associated summer warming caused the initial melt of the ice sheets, with feedbacks from ice and snow albedo changes (i.e., the melting of snow and ice exposed darker and more absorptive surfaces) and the carbon cycle further amplifying the warming. Orbital forcing continued to drive increases in boreal summer insolation until ~10,000 years ago, when it began the long-term decline toward present-day values.

Global annual average temperatures over the Holocene increased in response to the changes in orbital forcing and may have peaked during the mid-Holocene, ~9,000 to 6,000 years ago (Marcott et al., 2013). This was followed by a long-term cooling that lasted until just before the most recent warming trend, which began ~150 to 200 years ago with the advent of industrialization and anthropogenic greenhouse forcing. Other interpretations (Liu et al., 2014), however, suggest that this warming was much more centered in the Northern Hemisphere, that it increased continuously from the LGM to the preindustrial era, and that reconstructions showing a mid-Holocene peak in warmth followed by cooling are geographically biased toward the Northern Hemisphere. Regardless, this warming was the primary cause for the retreat of the ice sheets, remnants of which persisted well into the mid-Holocene. While they were extant, the ice sheets continued to affect weather and climate at regional and hemispheric scales through their direct influence on the atmosphere itself, through meltwater pulses of freshwater into the Atlantic, and through impacts on sea level and topography as the ice sheets melted and retreated.

Other Forcings and Feedbacks

Orbital forcing was the primary driver of the transition from the LGM to the Holocene, as well as the long-term (millennial-scale) temperature evolution of the Holocene itself. On shorter timescales (years to centuries), however, other sources of climate variability were also important. Energy output from the sun, for example, changes independently from orbital forcing and can be reconstructed from observations of sunspots and radiogenic isotopes in various proxies (Beer, 2000). Typically, high energy output from the sun is associated with a high number of sunspots and higher rates of production of these radiogenic isotopes. The most well-studied of these patterns of solar variability is the 11-year solar cycle, but the paleoclimate record also shows extended decadal to centennial periods of relatively quiescent solar activity. Two of the most notable recent periods of extended low solar output are the Maunder Minimum during the 17th and 18th centuries and the shorter Dalton Minimum in the early 1800s.

Volcanic eruptions have also had large impacts on Holocene climate, primarily through the emission of aerosol-forming particles that reflect and absorb solar radiation at upper levels in the atmosphere (Robock, 2000). Solar energy transmittance through the troposphere and energy absorption at the surface are both reduced following many eruptions, resulting in significant, but usually short-lived, cooling. The eruption of Mount Tambora in 1815 is particularly notable, as it caused widespread cooling across North America and Europe, resulting in growing season frosts and crop failures the following year (the year 1816 became "The Year Without a Summer"; Luterbacher & Pfister, 2015). Other eruptions with documented cooling effects include Laki (1783–1784), El Chichón (1982), and Pinatubo (1991). Normally, the effects of these eruptions persist for only a few years, after which the aerosols in the atmosphere are removed by precipitation or gravitational settling. In some instances, however, volcanic cooling may be stored as cold temperature anomalies in the oceans, allowing their impact to persist for decades or centuries. This is one hypothesis offered to explain the **Little Ice Age (LIA)** (Miller et al., 2012), which was not a true glacial period but rather an anomalous cold interval that began in the 14th and 15th centuries. According to this hypothesis, the LIA was triggered through a high number of volcanic eruptions that, along with feedbacks from the oceans and sea ice, induced a multi-century cool period that persisted until industrialization began in the early to mid-19th century. Other factors may have also been involved in the regional and global manifestation of the LIA, including low solar activity (e.g., from the Maunder and Dalton Minima) and regional ocean-atmosphere variability (Wanner et al., 2011).

GHG concentrations in the atmosphere increased across the transition from the LGM to the Holocene and over the course of the Holocene itself. During Pleistocene glacial intervals (including the LGM), atmospheric carbon dioxide (CO_2) concentrations averaged below 200 parts per million (ppm), much lower than the ~250 to 280 ppm typical for the Holocene and other interglacials. Higher CO_2 levels (likely sourced from the oceans) during the interglacials acted as a positive feedback, amplifying the initial warming driven by orbital forcing and retreat of the ice sheets. The Early Anthropogenic Hypothesis has also been invoked to explain preindustrial increases in CO_2 and methane (CH_4) that appear different compared to increases during previous interglacials (Ruddimann, 2007). This controversial hypothesis posits that Holocene increases in CO_2 and CH_4 concentrations can be partially explained by anthropogenic deforestation and the expansion of agriculture over the last 10,000 years. Beginning with the large-scale exploitation of fossil fuels in the 19th century, concentrations of many GHGs reached levels (e.g., more than 400 ppm of CO_2 at the time of this writing in spring 2018) that are likely unprecedented in the last 800,000 years.

Beyond changes in external climate forcings or feedbacks, modes of ocean-atmosphere variability relevant for hydroclimate have also shifted over the Holocene. There is some evidence that the general strength of El Niño Southern Oscillation (ENSO) *variability* was weaker during the mid-Holocene compared to the 20th century or earlier in the Holocene (Carré et al., 2014; Cobb et al., 2013), though the connection to orbital forcing, if any, is still uncertain (Emile-Geay et al., 2016). More recently, the ENSO system may have been biased toward more frequent La Niña events during the **Medieval Climate Anomaly** (MCA; ~800 to 1300 CE) (e.g., Coats et al., 2015), a possible explanation for some major North American drought events in the paleoclimate record (Coats et al., 2015; Seager et al., 2008). The Atlantic Ocean may have also been warmer during much of the MCA, with the Atlantic Multidecadal Oscillation shifted to a more positive phase (Coats & Smerdon, 2017; J. Wang et al., 2017), a situation that also would have favored enhanced drought over North America. Reconstructing these modes of variability and their teleconnections to terrestrial regions can be difficult, however. Marine proxies are widely available over the Holocene, but higher-resolution records that can resolve Common Era variability are much rarer. Reconstructing these modes of ocean variability for more recent centuries is thus more difficult and often based in part on terrestrial records and the associated teleconnections, which can be unstable (as discussed in chapter 3). Attributing hydroclimate events in the paleoclimate record to specific modes of ocean-atmosphere variability is therefore difficult, and any inferences made from the paleoclimate record must be treated with caution.

Hydroclimate in the Holocene

We divide our discussion of Holocene hydroclimate into two parts. In the first, we focus on large-scale hydroclimate shifts from the mid- to late Holocene, over which time orbital forcing and the retreat of the ice sheets played dominant roles. Even though these factors appear to be sufficient to explain some of the most important large-scale hydroclimate events, such as the millennial-scale wetting associated with the Green Sahara, there remain some significant uncertainties. These include the drivers of some important short-term events (e.g., the 4.2K event in the Middle East, discussed in the following section), rapid shifts inconsistent with more gradual forcing changes (e.g., the collapse of the Green Sahara), and the role of various mechanisms (e.g., land-atmosphere feedbacks). Following this, we pivot to examine regional hydroclimate changes over the last 2,000 years (the Common Era). Substantially more data, at much higher spatial and temporal resolution, are available over this period, allowing for much more detailed and precise spatiotemporal reconstructions of hydroclimate. Forcing changes during the Common Era (e.g., solar and volcanic) are weaker and relatively short-lived, however. This makes it more difficult to associate major hydroclimate events with specific forcings and points to a potentially more important role for internal variability and feedbacks during many of these events (e.g., tropical Pacific forcing of the MCA megadroughts).

The Mid- to Late Holocene (~9,000 to 2,000 Years Ago)

During the mid-Holocene, the orbitally forced peak in boreal summer insolation enhanced summer warming over the Northern Hemisphere. This caused a northward shift of the Intertropical Convergence Zone (ITCZ) and an intensification of the summer monsoons in the Northern Hemisphere, driving major and widespread shifts in hydroclimate. Much of the evidence for the hydroclimate changes during this interval comes from reconstructions of lake levels (figure 4.2) (Wanner et al., 2008), which, over much of the Northern Hemisphere, were higher and indicative of wetter conditions, especially in the major monsoon regions of Africa and Asia.

Some of the most dramatic wetting during this period occurred across the Sahara Desert, associated with the expansion of the West African monsoon. Evidence for this Green Sahara can be found in a variety of paleoclimate proxy records (figure 4.3), including reconstructed lake levels, outflow from rivers, and aeolian dust fluxes recorded in ocean sediments

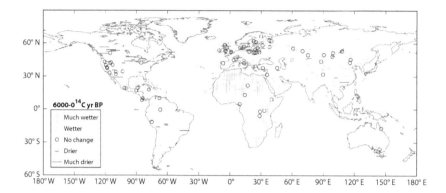

FIGURE 4.2 Differences in lake levels between 6,000 years before present (the end of the mid-Holocene) and the present. During this period, much of the Northern Hemisphere was wetter than today, a consequence of increased summer insolation shifting the ITCZ northward and strengthening the monsoons. *Source:* Figure 3, Wanner et al., 2008; data from the Global Lake Status Database.

(where wetter conditions result in less dust). Wetter conditions in Africa extended beyond the Sahara and also included much of East Africa and the Horn of Africa, a likely consequence of increased moisture transport from the Atlantic Ocean and warmer Indian Ocean temperatures (Tierney et al., 2011). The increased precipitation allowed for the expansion of vegetation and ecosystems across the Sahara, supporting animal life and facilitating the establishment of human settlements in areas that were too dry before and after the mid-Holocene. In the eastern Sahara, for example, settlements prior to this period were concentrated around the much more reliable water sources provided by the Nile River (Kuper & Kröpelin, 2006). Wetter conditions associated with the Green Sahara allowed for the establishment of settlements farther west, but these were eventually abandoned following the late Holocene weakening of the West African monsoon and onset of drier conditions more typical of the modern climate in the region (Manning & Timpson, 2014).

Whereas the existence and primary cause (orbital forcing) of the Green Sahara are relatively well constrained, there are significant outstanding questions regarding other contributing factors that remain to be solved. Foremost among them is the inability of many computer models to simulate the strength and extent of the enhanced monsoon during the Green Sahara

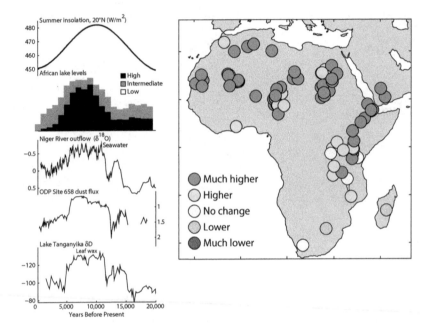

FIGURE 4.3 (*Left*) From 20,000 years before present to the present day, summer solar insolation at 20°N latitude and various proxy records of: African lake levels, Niger River outflow from oxygen isotopes of seawater, dust fluxes recorded in sediments at Ocean Drilling Program Site 658, and Lake Tanganyika lake levels reconstructed from isotopes in plant waxes. (*Right*) Difference between reconstructed lake levels for Africa during the mid-Holocene (9,000 years before; left panel) and the present levels. These data highlight the exceptional wetness that occurred during the mid-Holocene across North Africa. This wetting was likely caused by an intensification of the West African monsoon due to enhanced solar energy inputs during the summer along with feedbacks from vegetation and the land surface. Following the mid-Holocene, the region shifted to more arid conditions with the rapid collapse and retreat of the monsoon. *Source*: Jessica Tierney, adapted from Figure 2 and 3, deMenocal & Tierney, 2012.

and the speed of the collapse, which occurred much more quickly than would be predicted from changes in orbital forcing alone. Positive feedbacks from the expansion of vegetation across the Sahara are commonly invoked as one set of hypotheses (Claussen et al., 1999). The expansion of vegetation cover across the Sahara with the initial strengthening of the monsoon would have led to decreases in surface albedo and increases in evapotranspiration, further amplifying precipitation and the strength of the monsoon. Similarly, although the drying was likely initiated by orbitally driven declines in boreal summer insolation, the die-off of vegetation with the drying would have further weakened the monsoon, accelerated the drying, and hastened the monsoon collapse. In many models, however, even the inclusion of these vegetation feedbacks is insufficient to reproduce the full strength of the monsoon as recorded in the paleoclimate record. As a result, additional mechanisms have been invoked on top of the vegetation feedbacks, including changes in soil albedo (Levis et al., 2004), expanded forest cover in the Northern Hemisphere midlatitudes (Swann et al., 2014), and mineral dust aerosols (Tierney et al., 2017).

Isotope records from lakes and speleothems across the Mediterranean show wetter conditions during the mid-Holocene relative to the present day (Roberts et al., 2011). Peak wetness likely occurred between 6,000 and 5,400 years ago, at which point the region began a steady drying trend, reaching more present-day values by ~4,600 years ago (Finné et al., 2011). Notably, wet conditions during the mid-Holocene were larger and more coherent over the eastern Mediterranean than they were to the west. Unlike with the Green Sahara, however, available evidence suggests the mid-Holocene wetting in the Mediterranean occurred primarily during the winter, possibly due to a southward-shifted storm track. One notable event in the region occurred around 4,200 years ago and is thus referred to as the 4.2K event (deMenocal, 2001): a period of enhanced aridity in the eastern Mediterranean and Middle East that may have persisted for as long as 300 years. This was possibly a global event, and its occurrence is also recorded in proxy records in North America, India, and parts of Asia with a monsoon climate. Even though the cultural impact is still debated, there is evidence that it may have contributed to the collapse of the Old Kingdom in Egypt and the Akkadian Empire in the Middle East and may have caused major shifts in Indus Valley civilizations in India (Berkelhammer et al., 2012; deMenocal, 2001). The mechanism underlying the 4.2K event is still unclear, though hypotheses include a large volcanic eruption and changes in ocean circulation in the Atlantic Ocean (deMenocal, 2001).

As with Africa, there was a similar intensification of monsoon precipitation across Asia during the mid-Holocene, including in the areas of monsoon climate in India and East Asia and in the transitional region between the two (effectively Southeast Asia) (Li et al., 2014). Paleoclimate records show strong temporal coherence in the response across all three regions, with peak intensity occurring between ~10,500 and 5,500 years ago. The strong synchrony between the Indian and East Asian monsoon systems suggests that, much like in Africa, the monsoon intensification was at least partially caused by insolation forcing driving a poleward-shifted ITCZ in boreal summer. The decline in monsoon strength in the late Holocene, however, was much more gradual over Asia compared to the collapse of the Green Sahara in Africa, suggesting possibly weaker or nonexistent feedbacks from vegetation and the land surface.

Contrary to the nearly uniform wetting over much of Africa and Asia during the mid-Holocene, hydroclimate changes were more spatially heterogeneous over North America. Here, the Laurentide Ice Sheet (LIS) over Canada continued to exert a particularly strong influence on regional temperature and moisture patterns until ~6,000 years ago, affecting winter storm track positions and displacing the Atlantic Subtropical (Azores) High toward the equator. The mid-Holocene was also a time of warmer sea-surface temperatures in the subtropics, which may have influenced moisture advection into North America, as well as the local position of the Atlantic ITCZ. Central and northeastern North America during the mid-Holocene was relatively dry and warm, shifting to progressively cooler and wetter conditions from 6,000 years ago to the present day (Shuman & Marsicek, 2016). On millennial timescales, temperature and moisture are anticorrelated over this region, with warm (cool) conditions associated with a drier (wetter) overall hydroclimate. California and Nevada were wetter during the early to mid-Holocene, with drying in the region starting after 8,200 years ago as winter storm tracks moved north with the decline of the LIS (Steponaitis et al., 2015). Much like other Northern Hemisphere monsoon regions, the North America monsoon region in the southwestern United States and northwestern Mexico was stronger and wetter 9,000 years ago compared to today (Metcalfe et al., 2015). By 6,000 years ago, much of the U.S. Southwest began to dry out, and southern Mexico became wetter, changes consistent with the summer insolation decline, weakening of the North American monsoon, and southward retreat of the ITCZ (Metcalfe et al., 2015).

The terrestrial paleoclimate record is sparser in the Southern Hemisphere, and there is thus greater uncertainty regarding hydroclimate shifts and the associated causes during the Holocene. The midlatitudes in South America were drier during the mid-Holocene, from ~7,700 to 5,300 years ago, relative to today (Behling, 1995; Behling & Pillar, 2007; Fritz et al., 2001; Lamy et al., 2001). Notably, the Altiplano, a high plateau encompassing parts of modern-day Peru, Chile, and Bolivia, experienced multimillennial lows in lake levels from the mid-Holocene until ~4,500 years ago (Wanner et al., 2008), eventually reaching modern levels by ~3,500 years ago (Grosjean et al., 2003; Nuñez et al., 2002). Northern Australia likely reached its current levels of precipitation sometime after the end of the mid-Holocene with the onset of enhanced Walker circulation (Shulmeister, 1999). Signals over southern Africa are mixed, with some regions drier and others wetter during the mid-Holocene relative to today (deMenocal & Tierney, 2012).

Common Era (2,000 Years Ago to the Present)

The increased availability of annually to decadally resolved climate proxies (e.g., tree rings, corals, and ice cores), historical records, and archaeological data all allow for more highly resolved estimates of paleoclimate during the Common Era. This period marks a shift toward a reduced role for orbital changes and the increased prominence of shorter-term forcings (e.g., volcanic eruptions and solar variability) and internal variability (e.g., ENSO). Human activities involving land-use change and the burning of fossil fuels become globally important as well, especially with the onset of industrialization in the early 19th century.

At the global and hemispheric scales, the two climate events that stand out most during the Common Era are the MCA (~800 to 1300) and the LIA (~1300 to 1800). The exact dates are only coarsely defined, but these intervals can be seen clearly in various hemispheric and global temperature reconstructions of the last 2,000 years (Masson-Delmotte et al., 2013). The MCA was very likely warmer than the LIA at global and hemispheric scales, and temperatures in some regions may have even approached the warmth of the late 20th century. This was a time when conditions were warm enough for the Norse to establish settlements on Greenland, only to abandon them with the onset of colder conditions at the start of the LIA (Kintisch, 2016). The LIA, which may have been triggered by a fortuitous combination of high volcanic activity and lower solar output (the Maunder and Dalton Minima), was likely substantially cooler than the mid-20th century, resulting in significant glacial advances in many alpine regions. As with estimates of mid-Holocene temperatures, however, proxy information on temperature changes over the last two millennia is strongly biased toward the Northern Hemisphere, especially the Atlantic sector (North America and Europe). Further, in the absence of any singular, large-scale forcing (e.g., a mid-Holocene boreal summer insolation maximum), significant uncertainties remain about the causes of the MCA and LIA and even about the timing, extent, and coherence of the associated climate anomalies across regions and seasons.

North America is one of the most well documented regions for Common Era hydroclimate. This has been largely facilitated by the abundance of high-resolution, drought-sensitive proxy records across the continent, especially from lakes and tree rings. One of the most startling discoveries over this region has been the regular occurrence of **megadroughts**, multidecadal- to centennial-scale periods of drought during the last 2,000 years (B. Cook, Cook et al., 2016; Stine, 1994; Woodhouse & Overpeck, 1998). These events are defined primarily by their significantly greater persistence compared to most droughts in the more recent historical record, such as the Dust Bowl of the 1930s or the 1950s drought in the U.S. Southwest. Megadroughts have been documented in a variety of proxies and regions (Woodhouse & Overpeck, 1998), including the southwestern United States, Mexico, California, the Pacific Northwest, and the central plains. When compared to drought variability over the last two centuries, the megadroughts clearly stand out in terms of their persistence and, in some cases, their severity (figure 4.4). Another notable characteristic of the megadroughts is their temporal clustering. Most megadroughts occurred during the MCA, with the 12th and 13th centuries standing out as periods of especially widespread aridity across much of western North America. The last major megadrought was a late 16th-century event that affected much of the U.S. Southwest and central plains, after which drought variability in most regions shifted to patterns similar to those of the modern day. There is no clear consensus on what caused the megadroughts, why these events were apparently clustered during the MCA, or why they have not recurred in the last ~400 years (B. Cook, Cook et al., 2016). Megadroughts occurred across North America, affecting regions with often very different local climatologies,

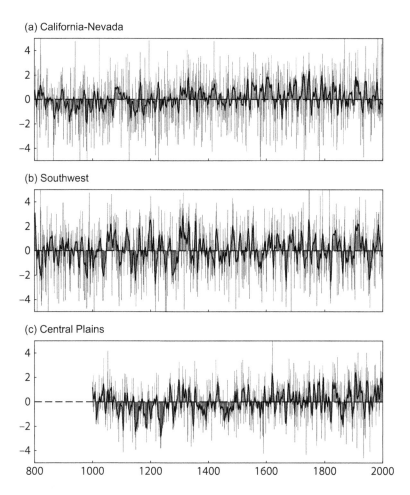

FIGURE 4.4 Tree-ring-reconstructed soil moisture variability (Palmer Drought Severity Index) for the three epicenters of megadrought activity in North America: California–Nevada (32°N–42°N, 125°W–114°W), the southwestern United States (28°N–38°N, 114°W–103°W), and the central plains (33°N–45°N, 103°W–90°W). All three regions experienced significant, multidecadal megadroughts during the MCA, especially during the exceptionally dry 1100s and 1200s. The notable persistence and severity of these events are clear when compared to more modest droughts in the record during the last several hundred years. *Source*: Data from the North American Drought Atlas, http://drought.memphis .edu/NADA/ (E. Cook, Seager et al., 2010).

hydroclimate regimes, and teleconnections to modes of ocean variability. As such, it is unlikely that a single, unifying mechanism could explain all aspects of megadroughts for all regions. Current evidence does, however, favor certain hypotheses for at least some megadroughts.

The tendency for megadroughts to cluster in time during the MCA has been used to suggest a role for exogenous forcing, possibly from increased insolation due to the paucity of volcanic eruptions and the modestly enhanced solar activity at the time. The mechanism most commonly invoked is that the positive radiative forcing during the MCA would have favored ocean conditions (i.e., a cold eastern tropical Pacific and a warmer Atlantic) conducive to drought across North America. Indeed, there is some evidence that the eastern tropical Pacific was anomalously colder and the Atlantic anomalously warmer for at least part of the MCA and that these ocean conditions were coincident in time with megadroughts across the U.S. Southwest (Coats, Smerdon, Cook et al., 2016; Coats, Smerdon, Karnauskas et al., 2016). The preferred clustering of megadroughts during the MCA does appear to be a unique feature of the climate at the time, likely a consequence of a mean shift toward drier conditions unlikely to be explained by internal climate variability alone (Ault et al., 2018; Coats, Smerdon, Karnauskas et al., 2016). Ocean teleconnections to drought over the central plains are generally weaker, and, over this region, land surface feedbacks may have played an important role during the megadroughts (B. Cook et al., 2013). There is widespread geomorphological evidence for significant dune mobilization and aeolian sediment transport during the megadroughts across the central plains, a situation that could have occurred only with significant vegetation mortality and loss of the stabilizing influence of vegetation on the soils. Such a loss of vegetation would have increased the surface albedo (i.e., less energy would be available for convection and precipitation), shifted the energy budget to favor sensible over latent heating (i.e., less overall evapotranspiration would occur), and caused increased wind erosion and dust storm activity, all phenomena capable of further suppressing precipitation. Finally, it is likely that internal variability in the ocean-atmosphere system may have also been important. Indeed, megadrought-like periods are generated by such processes in many models, even in the absence of significant external forcing (Ault et al., 2018; Coats et al., 2015; Hunt, 2011; Stevenson et al., 2015).

At least two of these megadroughts have been connected to major shifts in indigenous populations in pre-Columbian North America. The first occurred in the late 1200s, centered in the southwestern United States. This drought, which lasted over 20 years, is believed to have contributed to the abandonment of the iconic cliff dwellings in the Four Corners region of the southwestern United States by the Ancestral Puebloans. This event, and its possible connection to the Puebloans, was first described in some of the earliest work using tree rings to study drought and archaeology in the U.S. Southwest (Douglass, 1929, 1935). Another megadrought occurred in the early to mid-1300s, centered in the Mississippi River valley. This drought is believed to have caused the abandonment and depopulation ~1350 of Cahokia and surrounding areas (Benson, Barry et al., 2007; Benson et al., 2009), which, at the time, constituted the largest urban center in North America north of Mexico.

Winters in southwestern North America and northern Mexico became cooler and wetter in the transition to the LIA, while eastern North America continued the long-term wetting trend that began earlier in the Holocene. Farther south, relatively wet conditions occurred during the MCA across much of southern Mexico, Central America, and the Caribbean, with drier conditions in the highlands of central Mexico and the Yucatán. One notable event in the region occurred during the MCA from ~800 to 1000. This was a multicentury period of below-average moisture availability across the region, documented primarily in cave and lake records across the Yucatán Peninsula (figure 4.5). From these records, it is estimated that annual precipitation may have declined, on average, by as much as 25–40 percent during these centuries (Medina-Elizalde & Rohling, 2012). This centennial-scale drought is believed to have contributed to the collapse of the Maya civilization and the abandonment of many of the urban centers throughout the region, a period known as the Terminal Classic Period of the Maya. In this case, broadly applying the umbrella term *collapse* likely obfuscates the complexity of the societal responses that occurred, including geographic differences in the timing of abandonment (Douglas et al., 2016). To date, the causes of this drought are still debated, with various arguments offered for internal variability, solar forcing, and even deforestation as contributing factors (B. Cook et al., 2012; Hodell et al., 2001; Hunt & Elliott, 2005; Oglesby et al., 2010).

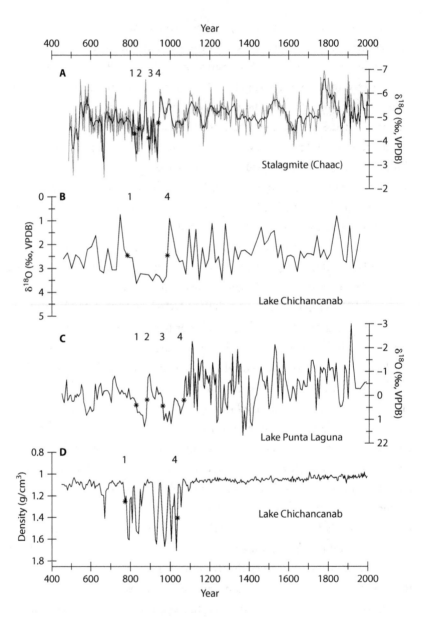

FIGURE 4.5 Hydroclimate proxies from one cave and three lake records from the Yucatán Peninsula, oriented so that the down direction on all four plots indicates drier conditions. All four records show a period of enhanced aridity between ~800 and 1000, a drought that coincided with the collapse of the Maya civilization. *Source*: Figure 2, Medina-Elizalde & Rohling, 2012.

Over Africa, studies of Common Era hydroclimate are hampered by a dearth of high-resolution proxies, such as annually dated tree rings and ice cores. There is evidence, however, that Africa continued its long-term drying from the mid-Holocene to the present, with significant decadal to centennial variability superimposed. One major event during the LIA, centered over coastal East Africa, was an extended pluvial from 1680 to 1765 (Tierney et al., 2013). This period of extended wet conditions likely coincided with anomalously cool sea surface temperatures in the eastern Indian Ocean. Following this pluvial, much of Africa experienced a multidecadal decline in rainfall, culminating in an episode of severe drought in the 1820s and 1830s that spanned nearly the entire continent (Nash et al., 2016). During this time, low water levels were recorded in Lakes Tanganyika and Rukwa, and famine afflicted parts of Tanzania and Kenya. In some regions (e.g., Lake Victoria), this period may have even been the driest of the last millennium. Several major droughts also occurred in the late 1800s in East Africa, including an event from 1888 to 1892 that caused a significant decline in the water level of Lake Tanganyika and a famine that may have killed one-third of the population of Ethiopia.

Over Asia, there is some evidence for an overall enhanced monsoon during the MCA and drier conditions during the LIA, but these subtle changes are overshadowed by much larger variability at shorter timescales and in specific regions. Central Asia and Mongolia, for example, experienced what was likely the wettest period of the last 1,100 years from 1211 to 1225, a pluvial that coincided with the expansion of the Mongol Empire from 1206 to 1227 (Pederson et al., 2014). Such wet conditions, combined with anomalous warmth at the time, would have increased plant productivity across the native grasslands, allowing the Mongols to support more domesticated livestock. This includes horses, which formed the backbone of their armies at the time, so it is believed that this era of favorable climate may have been integral to the rise and expansion of the Mongol Empire. In Southeast Asia, the Angkor civilization disappeared following a long period of weaker monsoon activity in the 14th and 15th centuries that was interspersed with some of the wettest years in the paleoclimate record (Buckley et al., 2010). In addition to the impact of the drought on crop yields, the rapid reversals between dry and wet conditions may have damaged their sophisticated water-management infrastructure, including canals and reservoirs, beyond repair. Several other major droughts are recorded in Asia in the tree-ring record and historical documents, all associated with major social changes (E. Cook, Anchukaitis et al., 2010). These include the drought in northern China that predated the fall of the Ming Dynasty in 1644, a mid-18th century drought ("Strange Parallels") that affected India and Southeast Asia concurrently, and a period of drought in the 1790s that caused severe famine in India. One of the most severe and widespread droughts occurred in 1876–1878, affecting India, Southeast Asia, and China. This event was part of a global pattern of climate disruption, caused in part by a strong El Niño event, that may have contributed to 30 million famine deaths worldwide (Davis, 2002).

For other regions, proxies during the Common Era are either sparse (e.g., in the Southern Hemisphere) or do not show fundamentally large or well-organized shifts in hydroclimate over the last 2,000 years. Europe cooled from the MCA into the LIA, but there is little evidence for large-scale coherent moisture changes. From 1314 to 1316, however, much of northern Europe was exceptionally cool and wet, conditions that led to large-scale crop failures and famine in an event referred to as the "Great Famine" (Rosen, 2014). In South America, drought conditions in Patagonia occurred at the same time as some of the worst megadroughts in California during the MCA (Stine, 1994), perhaps suggestive of some coherent response across hemispheres. In Australia, one of the earliest documented droughts occurred in the austral summer of 1791–1792, shortly after the establishment of some of the first European colonies. This "Settlement Drought" caused a partial failure of the wheat harvest in southeastern Australia and severely impacted the water supply to Sydney (Palmer et al., 2015; Russell, 2009). This event may have been a response to the same El Niño event that caused drought over South Asia the same year.

FIVE

Climate Change and Drought

C limate change will have a large influence on drought dynamics through its impact on precipitation, evapotranspiration, runoff, soil moisture, and snow. Shifts in the hydrologic cycle, however, will vary widely in both magnitude and sign across regions and seasons, precluding any general conclusions about climate change and drought at the global scale. Further, the ultimate impact of climate change on drought and water resources will depend on a variety of factors, including how droughts are defined, how human water use and demand shift in the future, uncertainties in climate models, future greenhouse gas (GHG) emissions, and our understanding of the climate system itself. In this chapter, we will review the current understanding of climate change and drought. How will climate change affect the processes controlling the water cycle and surface moisture budget? What processes will drive regional drought trends in the future, and how will this affect human water scarcity? What are the major uncertainties that still need to be resolved to have confidence in future drought projections? And can we already detect an impact of climate change on recent drought events?

Overview of Anthropogenic Climate Change and Climate Models

The paleoclimate record demonstrates how natural forcings, feedbacks, and even internal dynamics of the ocean-atmosphere system can drive significant shifts in global and regional climate. Over the next several hundred to several thousand years, however, the long-term evolution of the global climate system will be dominated by human activities, primarily because of increased atmospheric GHG concentrations. Of these gases, carbon dioxide (CO_2) is the single most important because fossil fuel burning and land-use change (e.g., deforestation) have dramatically increased its concentrations in the atmosphere over the last 200 years. This is illustrated most clearly in the observational record from the Mauna Loa Observatory in Hawaii (https://scripps.ucsd.edu/programs/keelingcurve/), which shows an increase in the global concentration of CO_2 from ~315 parts per million (ppm) in the 1950s to over 400 ppm in the present day. Modern-day concentrations in the atmosphere far exceed even the highest levels recorded in ice cores spanning the last 800,000 years, when CO_2 levels dropped below ~200 ppm during the cold glacial periods and rose to between 250 and 300 ppm during the warmer interglacials (CO_2 concentrations before industrialization in our own interglacial, the Holocene, were ~280 ppm). Modern CO_2 concentrations of >400 ppm in the atmosphere thus clearly stand out as extreme and far outside the range of these natural variations. In step with increasing GHG concentrations, global annual average temperatures have increased over the last ~150 years (figure 5.1), with especially rapid warming occurring since ~1970. Although figure 5.1 shows only one specific temperature record (GISTEMP, from the NASA Goddard Institute for Space Studies), other datasets (e.g., those produced by the National Oceanic and Atmospheric Administration and the Hadley Centre) show very similar results, lending high confidence to conclusions that Earth has warmed significantly.

But how do we confidently attribute this warming to anthropogenic activities rather than natural processes, such as the sun or volcanoes? Such conclusions rely primarily on climate models, computer programs that simulate the physics, chemistry, and biology of the climate system. Climate models are critical tools in the study of the climate system, are useful for studying

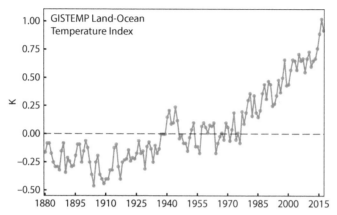

FIGURE 5.1 Global annual average temperature (K) anomalies (1880–2017, relative to a 1901–2000 baseline) from the GISTEMP Land-Ocean Temperature Index, produced by the NASA Goddard Institute for Space Studies (GISS). The long-term warming trend over the period covered by this dataset is clearly apparent. As of this writing, 2016 is the single warmest year in the record. *Source:* Data from GISTEMP, https://data.giss.nasa.gov/gistemp/.

past climate events and processes, and are the only tool available for generating projections of future climate states. They are also widely used in the science of *detection and attribution*, which seeks to determine the degree to which things have changed in the climate system (detection) and the likely cause of those changes (attribution), especially as it relates to anthropogenic climate change. An example for global temperature is shown in figure 5.2, which compares the evolution of global temperatures based on observations over the historical record with that based on model simulations with and without anthropogenic forcings. When run with both natural and anthropogenic forcings, climate models from the most recent generation (the Coupled Model Intercomparison Project version 5; CMIP5) and a previous generation (the Coupled Model Intercomparison Project version 3; CMIP3) reproduce the observed temperature trends with high fidelity. When the simulations use natural forcings and exclude anthropogenic forcings (including the long-term increase in atmospheric GHG concentrations), however, the models diverge substantially from observations and cannot reproduce the most recent warming. This demonstrates that recent temperature trends are best explained by anthropogenic changes to the climate system (primarily increased GHG emissions), providing some of the strongest evidence to date

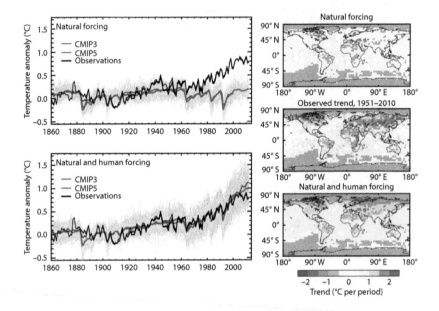

FIGURE 5.2 (*Left*) Time series of global and annual-averaged surface temperature change from 1860 to 2010. The top panel shows the results from two ensembles of climate models driven with just *natural climate forcings* (thin blue and yellow lines), the ensemble average temperature changes (thick blue and red lines), and observations (black lines). The bottom panel shows simulations by the same models but driven with both *natural forcings and human-induced changes in greenhouse gases, land use, and aerosols*. (*Right*) Spatial patterns of local surface temperature trends from 1951 to 2010. The top panel shows the pattern of trends from a large ensemble of Coupled Model Intercomparison Project version 5 (CMIP5) simulations driven with just natural forcings. The bottom panel shows trends from a corresponding ensemble of simulations driven with natural and human forcings. The middle panel shows the pattern of observed trends from the Hadley Centre/Climatic Research Unit gridded surface temperature dataset 4 (HadCRUT4) during this period. These analyses demonstrate that climate models can only reproduce observed warming trends when human forcings are included. *Source:* Reprinted from Bindoff et al., 2013, FAQ 10.1, fig. 1, p. 895.

that recent warming trends, and the associated changes in the climate system, are likely due to human activities.

Climate model simulations of the past can use observational estimates of important climate forcings, including observed increases in GHG concentrations, solar variability, and volcanic eruptions. To study climate change in the future, for which this information is not available, climate models are typically forced with scenarios of GHG concentrations and other anthropogenic forcings (e.g., land use) that extend to the end of the 21st century and beyond.

These scenarios are not intended to be policy recommendations or predictions. Rather, they are designed to represent a range of plausible futures based on various assumptions about future economic development, energy use (and sources), population growth, and other factors. The most recent generation of experiments (CMIP5) has focused on four of these scenarios, referred to as Representative Concentration Pathways (RCPs): low mitigation and strong warming (RCP 8.5), moderate warming and mitigation (RCP 4.5 and RCP 6), and aggressive mitigation and low warming (RCP 2.6). The associated number refers to the approximate radiative imbalance at the end of the 21st century (e.g., RCP 8.5 = +8.5 w/m^2). For this chapter, we will focus primarily on the high-warming scenario, RCP 8.5, to highlight the potential upper limit of changes that we might expect to see in the future.

Uncertainties in Future Climate Projections

Uncertainty is an integral part of all science. Climate models are no different, especially when used to estimate future climate states, and understanding these uncertainties is critical to determining the level of confidence we may have in projections for different variables, regions, and time intervals. Foremost among these are *process* uncertainties related to the simplified representation of often complex physics within the models. For some variables, such as precipitation, these processes cannot be explicitly simulated at the scale of the models and must be represented using simplified parameterizations. For others, such as evapotranspiration, we lack the observations necessary to properly evaluate the model simulations. And in some cases, we may not even possess a comprehensive understanding of the processes themselves.

A good example of the last case, with important ramifications for drought and climate, is the direct physiological response of vegetation to atmospheric CO_2 concentrations. At higher levels of atmospheric CO_2, plants can increase their *water-use efficiency*, reducing transpiration and surface water losses while maintaining, or even increasing, their carbon gain via photosynthesis. Because the bulk of terrestrial evapotranspiration occurs as transpiration, such an effect may be strong enough at the global level to counteract increased evaporative losses from higher temperatures, ameliorating projected increases in drought and aridity. Indeed, in most cases, vegetation in climate models will overwhelmingly increase water-use efficiency in response to increased CO_2 concentrations (Swann et al., 2016). However, real-world observations and experimental manipulations suggest that the ultimate response of vegetation to higher CO_2 concentrations is much more complex than the models suggest, with the magnitude of this effect varying sharply across species, ecosystems, and regions.

What is the best approach to deal with process uncertainties? Climate models often vary widely in their treatment of many different processes, and one can account for these uncertainties by analyzing responses across multiple models rather than focusing on the results of any specific model. Areas where the sign or magnitude of a change is consistent across models are generally viewed as the most *robust* projections. All models, for example, show widespread warming in all seasons in response to increasing GHG concentrations, whereas precipitation responses are typically less consistent, with strong differences across regions (see later in chapter). Significant agreement across models does not *necessarily* mean they are correct; rather, it suggests some convergence of evidence that, as always, must be evaluated in the context of observations and theory.

Another important source of uncertainty is natural climate variability, including modes like the El Niño Southern Oscillation, the Pacific Decadal Oscillation, and the Indian Ocean Dipole. This is because, over the next several decades, the variability associated with these modes, along with the random variability in the climate system itself, is likely to be larger than the GHG-forced trends in climate, making it difficult to confidently parse whether climate change is affecting recent and near-future drought and climate events. Because this variability is inherently random, it can act to either amplify or suppress anthropogenically forced climate trends. A clear example of this can be seen for the southwestern United States, a region that is expected to experience significant drying and increased drought risk with warming over the 21st century (B. Cook, Ault, & Smerdon, 2015; Seager et al., 2013). Seemingly in line with this expected climate change response, over the past 20 years the Southwest has experienced more-or-less continuous drought conditions. However, this has also been an extended period of cold sea surface temperatures in the eastern tropical Pacific (a high frequency of La Niña events), natural variability that has been linked to a recent slowdown in global warming and this persistent drought (Delworth et al., 2015). Therefore, it is difficult to attribute the recent drought in this specific region to climate change, even as we expect the region to continue to dry out with anthropogenic warming in the future. Indeed, internal climate variability, including shifts in the oceans and atmosphere unrelated to any specific climate mode (e.g., Malevich & Woodhouse, 2017; Wise, 2016) remains the dominant driver of most recent drought events and will continue to play an important role in the future, even in the face of anthropogenic climate change.

Finally, anthropogenic GHG forcing is expected to increase, and the signal-to-noise ratio of climate change to natural variability is expected to be much larger, by the end of the 21st century. On this time horizon, the largest uncertainties will come from the GHG emissions or concentrations scenarios, represented in the current generation of climate models by the RCPs. This is because the differences across these RCP trajectories are often significantly larger than the range covered by internal climate variability. Notably, these pathways represent some of the most difficult to constrain uncertainties in our understanding of the future climate because they depend on a multitude of political, social, and economic factors that are difficult, if not impossible, to model in any predictive sense.

Temperature

The most direct impact of increased GHG concentrations is seen in the change to the radiative balance at the top of the atmosphere, trapping more outgoing longwave radiation. In response, temperatures in the atmosphere and at the surface increase in order to achieve a new thermal equilibrium. Warming is therefore a more-or-less direct response to increased GHG concentrations, and temperature projections represent some of the most robust and consistent responses in the models (Collins et al., 2013; Knutti & Sedláček, 2013). The models show that, by the end of the 21st century, warming occurs in all regions and seasons relative to the modern-day or preindustrial era, amplified over land areas and at high northern latitudes (polar amplification), with the level of warming scaling with the forcing scenario. Compared to temperatures in the late 20th century (1986–2005), the global and annual average by the end of the 21st century (2081–2100) is projected to increase from 1°C at the low end (RCP 2.6) to over 4°C under high forcing (RCP 8.5) (figure 5.3). Separation between the various scenarios at 2100 is typically larger than the spread across models (shading) within a given scenario, highlighting the greater importance of the scenario (forcing) uncertainties over internal climate variability or model uncertainties by this time period. Generally, the magnitude of other climate changes and impacts in the models (e.g., sea-level rise, precipitation changes, and sea ice declines) scales with the magnitude of the forcing and warming. From an impacts perspective, this means there is an inherent trade-off between *mitigation* to reduce GHG emissions and warming and *adaptation* to deal with the ultimate impacts of climate change.

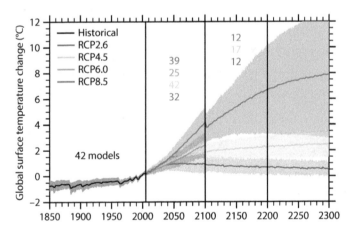

FIGURE 5.3 Time series of global annual mean surface air temperature anomalies (relative to a 1986–2005 baseline) from CMIP5 concentration-driven experiments. Projections are shown for each RCP for the multimodel mean (solid lines) and the 5–95 percent range (±1.64 standard deviations) across the distribution of individual models (shading). Discontinuities at 2100 are due to different numbers of models performing the extension runs beyond the 21st century and have no physical meaning. Only one ensemble member is used from each model, and numbers in the figure indicate the number of different models contributing to the different time periods. No ranges are given for the RCP 6.0 projections beyond 2100, as only two models are available. *Source:* Reprinted from Collins et al., 2013, fig. 12.5, p. 1054.

Humidity

One of the most direct drought-relevant impacts of warming involves humidity and the capacity of the atmosphere to hold water vapor. According to the Clausius-Clapeyron relationship (see chapter 1), for every degree of warming, the saturation vapor pressure in the atmosphere increases by ~7 percent. In areas with unlimited availability of moisture at the surface (e.g., the oceans), evaporation from the surface will respond quickly, and actual water vapor content of the atmosphere (i.e., specific humidity or actual vapor pressure) will increase to keep pace with the increased evaporative demand. Over the oceans and on longer timescales, this results in near-constant relative humidity (the ratio of actual vapor pressure to saturation vapor pressure) with warming. Over most land areas, however, relative humidity is expected to decline, and a related quantity, the vapor pressure deficit (VPD; defined as the difference between the saturation vapor pressure and the actual vapor pressure), is expected to increase (figure 5.4). This occurs because the supply of moisture to the atmosphere is limited over land by moisture availability at the surface and in the soils (A. Berg et al., 2016). Initially, evapotranspiration will typically keep pace with the increased demand, but as the surface runs out of water, evapotranspiration will eventually decline. Without this

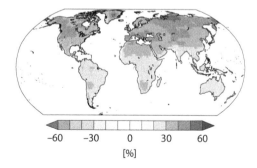

-60   -30   0   30   60
[%]

FIGURE 5.4 End-of-century changes (RCP 8.5 scenario, percent change from 1976–2005 to 2070–2099) in the vapor pressure deficit (VPD) for summer (June–July–August in the Northern Hemisphere; December–January–February in the Southern Hemisphere) from 17 CMIP5 models. The widespread and robust increase in VPD over land is a direct response to the robust warming in the models, a change that will likely exacerbate surface drying and drought impacts. Source: Justin Mankin.

additional input of moisture to the atmosphere, relative humidity will also decline, and VPD will increase. Here, terrestrial vegetation plays an important role because as soils dry out, plants will typically close their stomata to conserve moisture, further limiting evapotranspiration. As noted previously, higher levels of atmospheric CO_2 are also expected to increase the water-use efficiency of vegetation, further limiting evapotranspiration. The expected declines in relative humidity and increases in VPD with warming over many land regions are expected to exacerbate drought impacts and water stress. Warmer temperatures and higher VPD can amplify drying at the surface by increasing surface water losses, causing earlier snowmelt, and increasing the fraction of precipitation falling as rain instead of snow. High VPD is also strongly associated with increased forest mortality and stress (Williams et al., 2013), and recent trends toward increased wildfire have been attributed to increases in VPD from climate change over the western United States (Abatzoglou & Williams, 2016).

Precipitation

Global precipitation will increase in response to the global increases in evapotranspiration and water vapor from warming. However, the rate of precipitation increase per degree of warming is expected to be only about one-third of the humidity increase (only ~1 to 3 percent per degree of warming) because of additional energetic constraints on precipitation. This occurs because any increase in the upward latent heat flux to the atmosphere (evapotranspiration takes heat up from the surface; condensation and cloud formation add it to the atmosphere) must be balanced by radiative cooling of the troposphere.

A focus on global average increases in precipitation, however, masks important and often divergent precipitation trends that are readily apparent when looking at specific regions or seasons (figure 5.5) (Collins et al., 2013; Knutti & Sedláček, 2013). Regional changes in precipitation are dominated by warming-induced shifts in circulation ("dynamic") and water vapor ("thermodynamic"). At a broad level, and more so over oceans than the land, these processes result in an intensification of the global climatological pattern of precipitation, often referred to as the "wet get wetter and dry get drier" paradigm. Precipitation generally increases in the tropics and midlatitudes

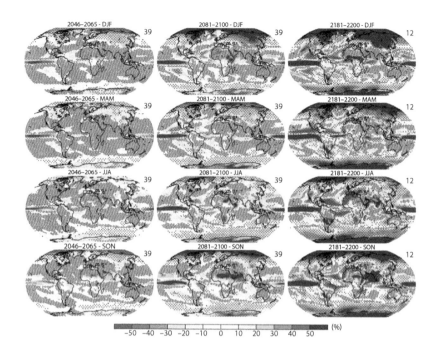

FIGURE 5.5 Multimodel CMIP5 average percentage change in seasonal mean precipitation relative to the reference period, 1986–2005, averaged over the periods 2045–2065, 2081–2100, and 2181–2200 under the RCP 8.5 forcing scenario. Hatching indicates regions where the multimodel mean change is less than one standard deviation of internal variability (least robust changes). Stippling indicates regions where the multimodel mean change is greater than two standard deviations of internal variability and where at least 90 percent of models agree on the sign of change (most robust changes). Source: Reprinted from Collins et al., 2013, fig. 12.22, p. 1078.

(areas of mean moisture convergence by the general circulation) and declines in the subtropics (areas of mean moisture divergence associated with the descending branches of the Hadley circulation). Precipitation projections, however, are generally less robust and coherent in the models relative to temperature projections and are prone to greater uncertainties. Much of this is related to the much more complicated suite of processes involved with precipitation in the models, including many that are parameterized rather than explicitly simulated (e.g., cloud physics). Note, as well, that regional changes over land often diverge substantially from the expected zonal average wet-get-wetter/dry-get-drier patterns, and, within regions, even the sign of expected changes can shift from season to season.

Land regions with the most robust precipitation declines at the end of the 21st century (areas denoted as *likely* to see decreased precipitation with warming) are the Mediterranean, the Caribbean, Central America, the southwestern United States and Mexico, and southern Africa. For these regions in the subtropics, these trends can be attributed primarily to changes in moisture divergence (the thermodynamic mechanism). There are also significant and robust increases in precipitation over the tropical oceans, in most major monsoon areas during the local summer, and in the mid- to high latitudes. The projected increases are especially strong and robust over high-latitude North America and Eurasia during boreal winter and occur mostly through the thermodynamic mechanism with additional contributions from a projected poleward retraction of the winter storm tracks. For other important regions of the tropics (e.g., the Amazon and West Africa), projected precipitation changes are not robust. One possible reason for this disagreement across models is the strong sensitivity of hydroclimate in these regions to nearby ocean temperatures (e.g., over the tropical Atlantic), which (depending on the model) may warm more or less than the global mean, affecting climate in nearby land regions. The apparent robust increase in precipitation over East Africa also likely reflects some pronounced deficiencies in the models. This is a region where it is notoriously difficult for many climate models to reproduce the observed seasonality and dynamics of precipitation, including teleconnections to various modes of ocean variability. Contrary to the models, the paleoclimate record strongly suggests that precipitation in East Africa will likely decline with warming (Tierney & Ummenhofer, 2015).

Beyond total precipitation, warming will affect the relative proportion of precipitation falling as rain versus snow, as well as snow dynamics at the land surface. As discussed previously, in colder regions and seasons (typically at high latitudes and elevations and during winter), a substantial fraction of cold season precipitation falls as snow rather than rain. This snow tends to accumulate at the surface over the course of the winter and melt rapidly in a single large pulse in the spring. This pulse can be critical for recharging soil moisture, streams, and surface reservoirs for the summer and growing season. With warming in the projections, however, the relative fraction of precipitation falling as rain increases at the expense of snow, even in areas where total precipitation is expected to increase (Krasting et al., 2013).

The largest declines in snowfall are expected in regions close to the 0°C isotherm in the midlatitudes, where projected warming is sufficient to push average temperatures above freezing. Farther north and at higher elevations (northern Canada, Russia, and the Tibetan Plateau), however, temperatures are projected to remain (on average) below freezing during the cold season, even with warming. In these regions, the increase in water vapor from the warmer atmosphere and the below-freezing temperatures are expected to result in overall increased snowfall and total precipitation.

With warmer temperatures and more precipitation falling as rain, snow stored at the surface is expected to melt earlier and more slowly (Musselman et al., 2017), distributing runoff more evenly across the year rather than in a concentrated pulse in the spring. Where water infrastructure is managed in such a way as to maximize capture of a large spring snowmelt pulse (e.g., California), this may lead to declines in reservoir storage if adjustments are not made to account for shifts in seasonal snow dynamics. Long-term observational records suggest, for many regions, that these changes in snowfall and snow at the surface are already occurring. Over the second half of the 20th century, for example, the western United States experienced significant declines of 15–30 percent in April 1 snowpack (Mote et al., 2018), an important indicator of snow water availability for the summer and growing season. This loss of snow represents ~25 to 50 km³ of water, similar in volume to Lake Mead, the largest and single most important human-made reservoir in the western United States.

Changes in precipitation are also expected at subseasonal and daily timescales (Polade et al., 2014). In response to increased humidity, the size and frequency of the most intense precipitation events are expected to increase. These types of precipitation events are the most efficient at generating runoff because they can quickly overcome the infiltration capacity of soils, and such a change could plausibly increase recharge of streamflow and surface reservoirs. Along with increases in the most extreme events, however, models respond by increasing the time between precipitation events (consecutive dry days). An increase in dry days and a lengthening of the time between precipitation events would make it easier for soils to significantly dry down between precipitation events, potentially exacerbating droughts and their impacts on ecosystems at the subseasonal scale.

Soil Moisture and Runoff

Water resources for people and ecosystems are most closely linked to surface hydrology through runoff (blue water) and soil moisture (green water). Understanding the response of these quantities to climate change is therefore imperative for quantifying linkages between climate change and drought impacts. Predicting soil moisture and runoff responses to climate change can be complicated, however, because they do not depend solely on changes in total precipitation. They are also affected, directly or indirectly, by other processes in the climate system (e.g., evaporative demand, snow dynamics, and precipitation intensity) and by the state of the land surface itself (e.g., vegetation cover and soil infiltration capacity). Increases in evaporative demand, largely driven by warming-induced increases in VPD, are expected to be especially important.

The importance of the added influence of warming and evaporative losses on soil moisture changes can be demonstrated using simplified drought indices, such as the Palmer Drought Severity Index (PDSI) and the Standardized Precipitation-Evapotranspiration Index (SPEI) (B. Cook, Smerdon et al., 2014). The PDSI and SPEI are simplified indicators of soil moisture availability, accounting for changes in moisture supply (precipitation) and demand (evapotranspiration, typically a function of temperature and VPD). Negative PDSI/SPEI values indicate drought conditions relative to some baseline, whereas positive values indicate wetter-than-normal conditions. As with many offline drought indices, PDSI and SPEI can be calculated using standard output from climate models (e.g., precipitation, temperature, and humidity), which allows the influence of specific processes or variables (e.g., precipitation versus evaporative demand) to be isolated. For example, when PDSI/SPEI for the end of the 21st century (2080–2099) is calculated under the RCP 8.5 scenario using *only* changes in precipitation (figure 5.6, middle panels), the results over land closely map to the precipitation projections discussed previously. Modest drying occurs in the Amazon, the Mediterranean, and various subtropical regions, with wetting over high northern latitudes and across much of monsoon Asia. When the same projections are made using *only* changes in potential evapotranspiration or evaporative demand (figure 5.6, bottom panels), widespread

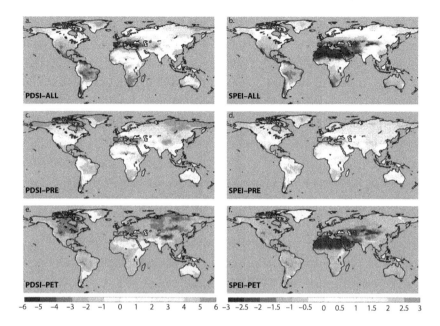

FIGURE 5.6 Ensemble average changes (RCP 8.5 scenario, 2080–2099 relative to a 1931–1990 baseline, 15 models) in the annual average Palmer Drought Severity Index (PDSI) and the 12-month Standardized Precipitation-Evapotranspiration Index (SPEI), forced with climate model projections, including changes in all variables (*top panels*, ALL), changes in precipitation only (*middle panels*, PRE), and changes in evaporative demand (e.g., temperature and humidity) only (*bottom panels*, PET). PDSI and SPEI are both indicators of soil moisture, and, when calculated using only precipitation changes, they show modest regional drying that is highly localized in areas with the most robust cross-model precipitation declines. The additional drying effect from a warmer atmosphere and increased evaporative demand intensifies the drying in these areas and spreads it to cover a much larger land area. Stippling indicates areas with robust model agreement regarding the sign of the changes. *Source*: Reproduced from Figure 11, B. Cook, Smerdon et al., 2014, https://link.springer.com/article /10.1007/s00382-014-2075-y.

and robust drying occurs across nearly all land areas. The combined effect of both processes (figure 5.6, top panels) thus amplifies drying in regions of major precipitation declines (e.g., the Mediterranean and southwestern North America) and shifts some areas significantly drier even in the face of negligible decreases, or even increases, in precipitation (e.g., northern Europe and the central plains in the United States). From a soil moisture perspective, PDSI indicates that evaporative demand will play an important role in long-term terrestrial hydroclimate trends.

As with many drought indices, however, PDSI and SPEI are relatively simplified representations of the land surface that do not explicitly simulate certain processes that may be important in these future drought projections. These include snow (all precipitation in PDSI/SPEI is treated as rain, regardless of the temperature) and direct CO_2 effects on water-use efficiency (changing atmospheric CO_2 concentrations have no direct effect on plant physiology and evaporative losses in PDSI/SPEI). Despite these deficiencies, however, these indices generally track changes in soil moisture from the more complex land surface models used in climate models with high fidelity (figure 5.7, top and middle panels). This is especially true for the near-surface soil moisture (the top 10 cm), which represents the water source likely to be most important for especially shallow rooted vegetation, including many crops. For soil moisture integrated across the full soil column (typically several meters), there are some differences and diminished drying over some regions compared to either PDSI/SPEI or the near-surface soil moisture. This divergence in soil moisture trends has been noted previously (A. Berg et al., 2017), but it is clear that the drying at depth still occurs over a much larger geographic area than would be predicted from projected precipitation changes alone. This strongly suggests that, at a broad level at least, PDSI and SPEI capture the most important processes affecting soil moisture in the climate model projections.

Changes in runoff are generally not as widespread, robust, or strong as those in soil moisture, but several regions do show some significant changes in the future (figure 5.7, bottom panel). Overall, runoff appears to be more sensitive to changes in precipitation than soil moisture is, with a much more limited influence of temperature and evaporative demand. Within the models, robust declines in runoff occur over southern and central Europe, the central United States, Mexico, and southern Africa. Robust increases in runoff occur across India and Southeast Asia and at higher latitudes in Canada and Asia. For many areas, changes in total runoff will also likely occur side-by-side with shifts in seasonal timing, especially for those regions where seasonal snowmelt is important.

The climate change projections discussed so far have focused on changes in the *mean* state—i.e., the long-term shifts in multidecadal average precipitation, soil moisture, runoff, and other variables. For many regions,

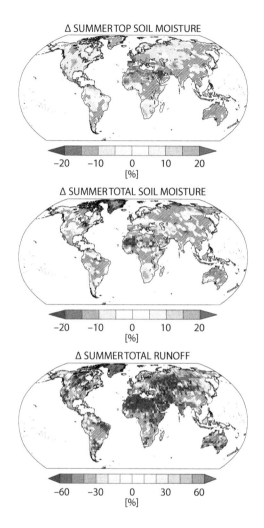

FIGURE 5.7 End-of-century changes (RCP 8.5 scenario, percent change from 1976–2005 to 2070–2099) in surface soil moisture (top 10 cm), full column soil moisture, and runoff for summer (June–July–August in the Northern Hemisphere; December–January–February in the Southern Hemisphere) from 17 CMIP5 models. Despite some differences, soil moisture responses taken directly from climate models map closely to results from offline drought indices, confirming the combined importance of changes in precipitation and evaporative demand. Broadly, however, runoff responses appear more sensitive to precipitation. Hatching indicates regions with inconsistent (nonrobust) responses across models. *Source:* Justin Mankin.

especially in the subtropics, the consensus among the models is that the future will be substantially drier *on average*. Changes in variability are less certain and robust in the models, but even modest shifts in the mean state can be sufficient to significantly increase the likelihood of occurrence (i.e., risk) of extreme drought events. Differentiating between drier *average* conditions and more frequent/intense/persistent extremes is important because the latter events are more likely to have larger impacts on ecosystems and people than are modest shifts in the mean state alone. With climate change, many regions are projected to experience increases in extreme drought risk.

As an example, soil moisture across the southwest and central plains of North America (evaluated using PDSI and soil moisture at depths of ~30 cm and ~2 to 3 m from the coupled climate models themselves) is expected to decline substantially with warming by the end of the 21st century (figure 5.8). Over the central plains, this occurs primarily through increased evapotranspiration (due to warming-induced increases in VPD), whereas over the southwest it is a consequence of both precipitation declines and increased evapotranspiration. Notably, the projections indicate a high likelihood that in the latter half of the 21st century (2050–2099), under the high-emissions RCP 8.5 scenario, the southwest and central plains not only will become drier than the historical period but also will likely be even drier than the most arid megadrought centuries in the paleoclimate record (1100–1300 CE).

Importantly, the shift in the mean state captured in these projections has profound implications for the risk of decadal (11-year) drought and multidecadal (35-year) megadrought events (figure 5.9). For the latter half of the 20th century (1950–2000), decadal drought risk in the models is relatively high (~40 to 60 percent), but multidecadal drought risk is low (less than 10 to 15 percent), a result largely consistent with observations. Persistent drought periods of a decade or less, for example, are a recurring feature over the North American central plains and southwest over the last 150 years, including famous historical events like the Dust Bowl drought of the 1930s and the 1950s drought. Notably, however, neither of these regions has experienced a megadrought for many centuries. During the latter half of

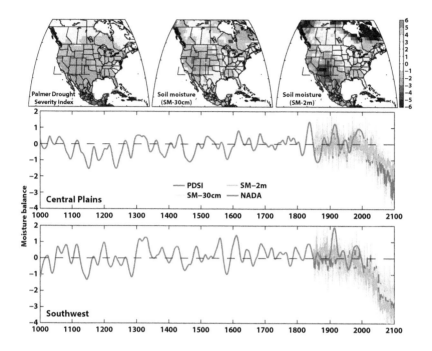

FIGURE 5.8 Multimodel projected changes (RCP 8.5 scenario, 2050–2099 relative to a 1931–1990 baseline, 17 models) in soil moisture for summer (June–July–August) for the southwest and central plains of western North America. The SM-30cm and SM-2m variables are scaled versions of soil moisture from the climate models integrated from the surface to ~30 cm and ~2 to 3 m depths, respectively. Brown lines represent 50-year smoothed versions of PDSI from a tree-ring reconstruction of drought from 1000 to 2005 CE. Although there are some differences across indicators, the aggregate multimodel response shows pronounced declines in soil moisture for both regions with warming over the 21st century. Notably, future average conditions will likely be substantially drier than the range of variability in the historical period in the models (1850–2005) and the full paleoclimate record for the last millennium. *Source:* Figure 1, B. Cook, Ault, & Smerdon, 2015, http://advances.sciencemag.org /content/1/1/e1400082.short.

the 21st century, however, all soil moisture indicators for both regions show substantial increases in the risk of both types of drought (over 80 percent in most cases), making these events a virtual certainty by the end of the 21st century under the RCP 8.5 trajectory. These results suggest, at least for western North America, the potential for a profound shift in hydroclimate and an associated increase in drought risk that is far outside recent historical or contemporary experiences.

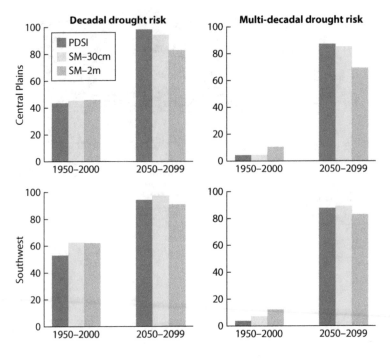

FIGURE 5.9 For the same multimodel ensemble used in figure 5.8, the estimated percentage likelihood (risk) of decadal drought (11-year) and multidecadal megadrought (35-year) occurrences for all three soil moisture drought indicators for the southwest and central plains of western North America. In the modern interval (1950–2000), decadal drought risk is modest, and multidecadal drought risk is minimal. By the end of the 21st century, under the RCP 8.5 scenario (2050–2099), the risk of both decadal and multidecadal droughts increases substantially, above 80 percent in most cases. *Source:* Figure 5, B. Cook, Ault, & Smerdon, 2015, http://advances.sciencemag.org/content/1/1/e1400082.short.

Climate Change, Water Scarcity, and Drought Exposure

Climatically driven increases in drought and aridity do not necessarily lead to increases in **water scarcity**, broadly defined as a lack of water resources available to meet the demands or needs of a population within a given region. Quantitatively, this is often expressed in terms of per capita water availability or the ratio of withdrawals to availability. Water scarcity depends not only on the availability of physical water resources but also on the requirements of a given population, requirements that may change significantly from region to region and over time. Water requirements for agriculture, for example, are typically much larger than those for municipal populations. Areas with extensive agricultural systems are thus more likely to experience water scarcity during drought events. A broad range of factors can affect water scarcity and drought resiliency, including municipal water-use efficiency and recycling, agricultural management (e.g., types of crops grown and use of irrigation), exploitation of groundwater, and capture and storage of water in artificial surface reservoirs. These factors depend heavily on complex social and political processes that are difficult to model and predict; thus, they represent major uncertainties in our understanding of what increased drought risk will mean from an impacts perspective.

One major factor that is likely to affect water scarcity in the context of climate change is population growth. One recent study (Smirnov et al., 2016) combined various scenarios of population growth and climate change to estimate drought exposure, defined as the population at any given time or location experiencing drought. Although not addressing the question of water scarcity specifically, such analyses provide some guidance on where drought impacts have the potential to be most significant. Even under a fixed climate scenario, a medium population growth scenario increases the global population exposure to drought by 35.5 million people because much of this population growth occurs in areas with significant natural drought variability. Alternatively, keeping population constant and using climate from the high-emissions RCP 8.5 scenario, drought exposure increases by 230 million people, reflecting the climate-driven increases in drought risk for many regions discussed previously. Population growth and climate change combined cause an even larger increase of 386.8 million people exposed to drought conditions. Thus, although recent evidence suggests that risks of exposure will depend primarily on climate, population growth in many of these aridifying areas will likely be an important secondary factor increasing drought exposure.

Climate Change and Recent Droughts

In this chapter, the discussion of climate change and drought has focused on projections into the far future, when anthropogenic GHG forcing will be stronger and the expected shifts in hydroclimate for many regions will be unequivocal. In recent years, however, many regions have experienced intense and, in some cases, unprecedented droughts, inspiring this question: Has climate change already begun to affect drought dynamics? This is an important and compelling question that nevertheless presents a significant challenge because, as noted previously, the noise of natural variability in the climate system is still large relative to the signal of climate change. The question is thus fundamentally much more complex than simply asking if climate change caused a given drought because any drought event will represent some mix of natural variability and anthropogenic climate change. We can, however, investigate characteristics of these events and attempt to determine if anthropogenic forcing played a role in their occurrence. The more appropriate questions are then: (1) Did climate change make a given drought event more likely to occur? (2) Has climate change made these drought events worse than they otherwise would have been? Here, we review some research on two of the most recent and important droughts in the early 21st century: those in Syria and California.

SYRIA

From 2007 to 2010, Syria experienced the single worst 3-year drought in the historical record (Trigo et al., 2010). This event caused declines of 47 percent in wheat yields and 67 percent in barley yields, crop failures that affected an estimated 1.3 million people (Gleick, 2014). These impacts were likely exacerbated by government policies that encouraged the overexploitation of land and water resources, increasing vulnerability to droughts (Gleick, 2014; Kelley et al., 2015). In response, an estimated 1.5 million people migrated from Syria's rural areas to its urban centers, possibly contributing to the uprising that would lead to the ongoing civil war in the country (Gleick, 2014; Kelley et al., 2015). Just from a climatological perspective, however, the drought itself appears as an exceptional event. Did climate change affect the Syrian drought?

Trends in precipitation over the Mediterranean region, which includes Syria, suggest that the predicted global warming–induced precipitation declines are already emerging. Cold season (November–April) precipitation across the entire Mediterranean has declined by 6.8 percent in recent decades

(1971–2010 versus 1902–1970) (Hoerling et al., 2012). This drying trend in precipitation is produced only in model simulations that include observed historical increases in anthropogenic GHGs (along with other natural and anthropogenic forcings), strongly suggesting it is forced by climate change.

The drought in Syria itself was part of a larger drought that extended over the eastern Mediterranean and Levant region (the geographic area including Israel, Palestine, Lebanon, Syria, and Jordan) from 1998 to 2012. From a paleoclimate perspective, this event was especially severe compared to the last ~900 years of natural variability, based on tree ring reconstructions of PDSI (B. Cook, Anchukaitis et al., 2016). The drought anomaly in 2000 over the Levant region stands as nominally the single driest year back to 1100 CE, and it is likely (has an 89 percent likelihood) that the multiyear drought from 1998 to 2012 was the driest such period of the past 900 years and also very likely (has a 98 percent likelihood) the driest period of the past 500 years. A more formal attribution study specifically focused on the Syrian drought was conducted by Kelley et al. (2015). In this study, the authors compared (1) precipitation observations versus observations where the climate change signal was statistically removed and (2) climate model simulations with natural and anthropogenic forcings versus simulations with only natural forcings. For the natural case (detrended observations and model simulations with natural forcings only), precipitation distributions in the region are significantly wetter than for the scenario that also includes an anthropogenic climate change signal. This overall dry shift in response to climate change, although modest, represents a two- to threefold increase in the likelihood of the most extreme observed droughts in the region. Combined with analyses of precipitation trends and the paleoclimate record, the evidence in aggregate strongly supports the case that the recent drought in Syria and the Levant was made more intense and more likely to occur because of anthropogenic climate change.

CALIFORNIA

The most recent major drought in California began in 2011–2012 and continued through 2016 (Diffenbaugh et al., 2015; S. Wang et al., 2017), manifesting as severe deficits in precipitation (Savtchenko et al., 2015), snowpack (Mote et al., 2016), surface reservoir storage (He et al., 2017), and groundwater (Xiao et al., 2017). Across models, however, there is large disagreement in the sign of projected precipitation trends over California in response to climate change, and the trends themselves are small relative to the large interannual variability over the state (N. Berg & Hall, 2015). As with most land areas, however, temperatures in California have warmed significantly over the last century in response to greenhouse warming. Can a climate change signal related to this warming be detected in the California drought?

Drought reconstructions from the paleoclimate record strongly suggest that some additional drying from anthropogenic warming occurred during the recent California drought. In figure 5.10, tree-ring-based reconstructions

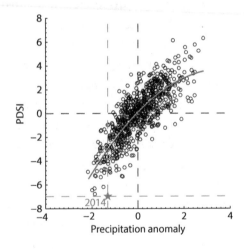

FIGURE 5.10 Comparisons between tree-ring-reconstructed summer (June–July–August) soil moisture (PDSI) and October–June precipitation (normalized) for California from 1293–2014; 2014, the most recent California drought year at the time of publication, is highlighted. From a precipitation perspective, 2014 was dry but well within the bounds of the reconstructed climate variability. However, 2014 was the single driest year in the reconstruction from a PDSI perspective, Further, PDSI for 2014 was significantly lower than would be predicted from the precipitation deficits alone. Overall, this evidence strongly suggests that warming-induced increases in evapotranspiration likely amplified the recent California drought. *Source:* Daniel Griffin and Kevin Anchukaitis, adapted from Figure 4, Griffin & Anchukaitis, 2014, https://agupubs.onlinelibrary.wiley.com/doi/full/10.1002/2014GL062433.

of meteorological drought (October–June precipitation) and soil moisture drought (PDSI) are compared from 1293 to 2014 (Griffin & Anchukaitis, 2014). The two are strongly and positively correlated: historically, drought variability in California (even in soil moisture) is driven primarily by variability in moisture supply from precipitation. For the worst year of the drought at the time of this study (2014), precipitation was low but not record breaking in the reconstruction, consistent with other analyses suggesting that precipitation deficits during the drought were mostly driven by natural variability (Seager et al., 2015). However, soil moisture for 2014 is the lowest in the reconstruction and also significantly lower than would be predicted from the precipitation anomaly alone. This suggests an important secondary role for warming temperatures and increased evaporative losses in amplifying the soil moisture drought. The impact of this warming is also reflected in reconstructions of spring snow water equivalents for the Sierra Nevada, which in 2015 were likely the lowest of the last 500 years (figure 4.1).

This impact of warming on soil moisture and snow is supported by other studies. Williams and colleagues (2015) quantified the California drought using PDSI, recalculating the magnitude of this event with the anthropogenic warming trend, and its impact on evaporative demand, removed. By comparing these estimates to PDSI calculations that include the underlying warming trends, the relative contribution of anthropogenic warming to drought severity can be estimated. In all cases, the severity of the multiyear (2012–2014) drought and its single worst year (2014) increased when anthropogenic warming was included. In the 2012–2014 mean, it was estimated that 8–27 percent of the underlying drought anomaly could be attributed to this climate change effect, whereas for 2014 alone the contribution was 5–18 percent. The impact of warming on snow is also strongly supported. Consistent with the previously highlighted paleoclimate reconstruction, spring snow cover in the Sierra Nevada was at a record low according to both surface observations (Mote et al., 2016) and satellite observations (Margulis et al., 2016). The impact of warming on this snow drought was also explicitly investigated using climate model simulations (N. Berg & Hall, 2017); the results demonstrated that anthropogenic warming reduced the overall snowpack by ~25 percent and snowpack at lower elevations by as much as 43 percent. Thus, while the primary driver of the California drought (precipitation) was likely dominated by natural variability, warming from climate change intensified the drying in both soil moisture and snow during this event.

SIX

Case Studies

The Dust Bowl and Sahel Droughts

Among the many persistent droughts that occurred during the 20th century, two events are notable for their intensity, their impacts, and the interest they generated in the scientific community. The first is the Dust Bowl of the 1930s, a nearly decadal long drought that afflicted much of the central plains in North America. Combined with a major economic depression and poor land-use practices, this drought became one of the worst natural disasters in U.S. history, resulting in historically high levels of land degradation, human migration, and dust storm activity. The second is a persistent drought that affected the Sahel region of West Africa from the late 1960s through the early 1990s. This multidecadal period of below-average rainfall ravaged ecosystems and subsistence farming communities in the region, inspiring new schools of thought and investigations into drought, desertification, and climate change. Here, we review these major droughts, discussing their causes and impacts (including the role of human activities) and some of the major lessons learned from studying these events.

The Dust Bowl

The term *Dust Bowl* was coined in 1935 by Robert E. Geiger after Black Sunday, a massive dust storm that blew across the Oklahoma Panhandle on April 14, 1935. Geiger, an Associated Press reporter, experienced the storm in Boise City, Oklahoma, and afterward would write, "Three little words achingly familiar on a Western farmer's tongue rule life in the dust bowl of the continent—if it rains" (Worster, 2004). Originally, the term was used to refer to the specific area of the southern plains near the Oklahoma and Texas Panhandles that was the center of the most intense dust storm activity and wind erosion during the drought. Later its use would expand, and *Dust Bowl* now refers more broadly to the entirety of the drought that affected most of the central plains from 1931 to 1940.

AGRICULTURE IN THE CENTRAL PLAINS

The roots of the Dust Bowl began with the movement of agricultural production from the eastern United States to the central plains beginning in the late 1800s (Lee & Gill, 2015) (figure 6.1). Forced removal, displacement, and genocidal actions against the indigenous tribes opened land for settlement and allowed for the westward migration of the U.S. population. Railroad companies were responsible for much of the initial push because they needed to create demand for travel services to the region, and most expansion took the form of ranching and dryland (nonirrigated) farming. The expansion of agriculture into regions previously thought to be too dry to farm (e.g., west of the 100th meridian) was further encouraged by various ideas promulgated in the media. This included "Rain follows the plow," the mostly erroneous notion that farming would increase local rainfall and thus the suitability of relatively dry regions for agriculture.

Agricultural production continued to intensify and expand over the central plains in the 1910s and 1920s for three primary reasons (Lee & Gill 2015). First, much of the early 20th century in Western North America, including the Central Plains, was anomalously wet, creating relatively favorable conditions for growing crops. Second, the introduction of mechanization, such as tractors and combines, made it easier to put more land into production. Third, wheat prices were relatively high over much of this period, mostly due to increased demand during World War I, creating strong economic incentives to boost productivity. The native grasslands, dominated by perennial species well adapted to drought, were thus progressively

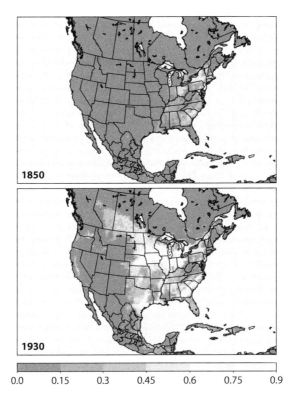

FIGURE 6.1 Cropland distribution (fractional area) over North America in 1850 (*top*) and 1930 (*bottom*). From the late 19th century through the mid-20th century, agricultural production in the United States shifted from the eastern states into the Midwest and central plains. This expansion replaced much of the native, drought-resistant grasslands across the region, setting the stage for large-scale land degradation with the onset of the Dust Bowl drought in the 1930s. *Source*: Data from Historical Croplands Dataset, 1700–1992, from Ramankutty & Foley, 1999, https://nelson.wisc.edu/sage/data-and-models/datasets.php.

replaced with highly managed, drought-sensitive, annually harvested crops (wheat) during a period when few major droughts occurred. Beginning in the 1920s, crop prices dropped as production increased and the demand from World War I abated. And although farmers were managing for water, few were concerned about erosion control. Lacking techniques for soil conservation and experience with significant droughts in the region, farmers were therefore ill prepared to deal with the onset of drought conditions in the 1930s.

CAUSES OF THE DROUGHT

The Dust Bowl drought varied in severity from year to year, with the most widespread and intense drying occurring in 1931, 1934, and 1936 (figure 6.2). Drought conditions were especially widespread in 1934, with evidence from the paleoclimate record suggesting that this was the worst single drought year of the last 1,000 years in terms of intensity and extent (B. Cook, Seager, & Smerdon, 2014). The multiyear drought was particularly severe in the southern

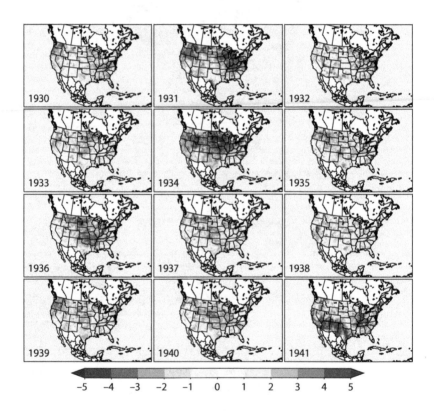

FIGURE 6.2 Summer soil moisture (June-July-August, Palmer Drought Severity Index, anomalies relative to a 1901-2000 baseline) from 1931-1941. Drought anomalies during the Dust Bowl of the 1930s were most widespread and intense in 1931, 1934, and 1936, with dry conditions in most years centered in the central plains and Midwest. The dominant pattern of ocean forcing during much of the Dust Bowl was a modestly cold eastern tropical Pacific (La Niña) and a warm tropical Atlantic, conditions conducive to drought across western North America. The drought itself ended, in part, as a consequence of a large El Niño event in 1940 and 1941. *Source:* Data provided by A. Park Williams, Williams et al., 2015.

plains (Oklahoma and Texas) and northern plains (Nebraska, North and South Dakota, and Montana).

The primary, or at least initial, cause of the drought was likely the pattern of anomalous sea surface temperatures (SSTs) in the tropics during the 1930s, first investigated in detail by Schubert et al. (2004). During the 1930s, the eastern tropical Pacific was anomalously cool, with a relatively high frequency of La Niña events, whereas the tropical Atlantic was warm, conditions generally conducive to drought over North America. Schubert et al. demonstrated that when an ensemble of climate model simulations is forced with these SST boundary conditions, the models produce the broad features of the precipitation deficits associated with the drought. The most intense precipitation deficits during the Dust Bowl occurred during the summer and fall, and the model simulations further reproduce this seasonality of the drought. Schubert and colleagues (2004) also identified a critical role for the land surface, using an alternative set of model experiments where the researchers disabled feedbacks in the model between the land surface (i.e., soil moisture) and the atmosphere (i.e., precipitation). Without these feedbacks, model precipitation deficits were significantly smaller than observed, highlighting the importance of land-atmosphere interactions for amplifying the initial SST-forced precipitation deficits. These model experiments, however, still had some difficulty capturing many of the important details of the Dust Bowl drought. For example, the models overestimated the drying over the U.S. Southwest and Mexico and underestimated the spatial extent of the drying over the northern plains and Midwest. Further, the model ensemble, even with full land-atmosphere coupling, underestimated the magnitude of the precipitation deficits in the core Dust Bowl region.

Despite the importance of SST forcing during the Dust Bowl (further investigated in Seager et al., 2008b), it is likely that other factors also played important roles in making this drought truly exceptional. Hoerling and coauthors (2009) highlighted how extreme internal atmospheric variability was required to generate the observed intensity of drought over the northern plains, where SST teleconnections are typically weaker. Focusing on 1934, B. Cook, Seager, and Smerdon (2014) also noted that intense ridging over the Pacific Coast and central plains during this widespread drought year was inconsistent with what would be expected from the La Niña that winter, again pointing to the importance of internal atmospheric variability. More than anything else, though, the Dust Bowl was defined by the widespread land degradation that occurred at the time. And there is evidence that the impact of the associated land surface changes on the atmosphere was critical for amplifying precipitation deficits and heat extremes during this event.

LAND DEGRADATION DURING THE DUST BOWL

One of the defining features of the Dust Bowl, one that differentiates this event from other 20th-century droughts in the United States (e.g., the droughts in the Southwest in the 1950s and at the turn of the 21st century) is the scale of land degradation that occurred. The degradation is rooted primarily in the widespread crop failures caused by the drought across the central plains, the magnitude of which can be seen clearly in historical wheat harvest records, expressed as the fraction of the area planted that was actually harvested (figure 6.3). From 1921 to 1930, prior to the Dust Bowl, the area planted in wheat in the United States that generated a harvest averaged 90.8 percent. During the worst years of the Dust Bowl (1933–1937), this dropped to an average of 71.8 percent per year, with the worst harvests in 1934 (67.7 percent) and 1936 (66.4 percent). This level of crop failure was substantially

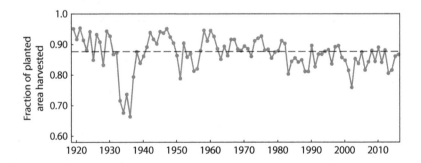

FIGURE 6.3 Annual wheat harvest in the United States, expressed as the fraction of the planted area harvested every year from 1919 to 2016. The median harvest fraction (~0.877) is shown by the dashed horizontal line. Harvests during the Dust Bowl drought stand out as particularly low, especially during 1934 and 1936. This highlights the unprecedented level of crop failure during this drought, especially when compared to subsequent droughts in the 1950s and early 21st century. Source: USDA Agricultural Baseline Database, https://www.ers.usda.gov/data-products/agricultural-baseline-database/.

worse than that during any subsequent event, including during the 1950s drought, when the single worst harvest was 78.9 percent in 1951.

Prior to the Dust Bowl, large dust storms were mostly associated with prairie fires, events that often occurred during droughts in the 1800s (Lee & Gill, 2015). The unprecedented loss of crops and natural vegetation during the Dust Bowl, however, caused significant declines in natural erosion control services (Lee & Gill, 2015). Both natural vegetation and healthy, growing crops reduce the wind speed and protect soils from wind erosion. Additionally, native grasses typically have dense root systems extending to significant soil depths and return substantial quantities of organic matter to the soil, both features that increase the water-holding, binding, and erosion-resisting capacities of soils. Crops, by contrast, typically have much shallower roots, and, because they are harvested, they do not return the same organic matter to the soils, leaving soils exposed for extended periods (Lee & Gill, 2015).

Both the replacement of native grasslands with croplands and the widespread failure of crops to grow during the drought increased the vulnerability of the landscape to erosion and degradation during the Dust Bowl. Poor land-use practices also contributed; for example, farmers commonly used the one-way disc plow, which exacerbated erosion susceptibility by breaking up the soil and making it difficult to maintain crop residue (a way to control erosion) on the surface (Lee & Gill, 2015). Facing the financial losses from the crop failures and the general economic depression, many farmers were forced to abandon their farms because they could not afford to replant after failed harvests. The abandoned farms were often left barren because few operators were interested in or capable of taking over, further exacerbating erosion issues (Lee & Gill, 2015) and also causing damage to neighboring farms when eroded soil was deposited on nearby crops and fields (Hansen & Libecap, 2004).

Significant levels of wind and water erosion affected nearly all of the central plains during the Dust Bowl (figure 6.4). Quantitative estimates are difficult to constrain with accuracy, however, because large-scale surveys of erosion and topsoil losses did not begin until the 1930s. Many estimates are therefore based on cumulative erosion damage after the Dust Bowl drought ended (Hornbeck, 2012). The worst erosion (characterized as high erosion, meaning more than 75 percent of the topsoil was lost) was concentrated in

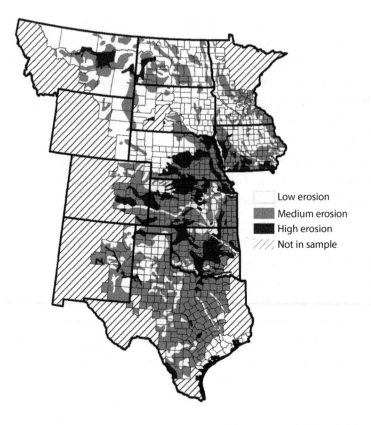

FIGURE 6.4 Cumulative erosion damage for the western United States assessed at the end of the Dust Bowl: low (less than 25 percent of topsoil lost), medium (25–75 percent of topsoil lost), and high (more than 75 percent of topsoil lost). The worst damage was in the titular Dust Bowl region of the southern plains (Oklahoma, Kansas, and Nebraska), central plains, eastern Montana, and western North Dakota. *Source:* Data from National Archives in College Park, MD, RG 114, Cartographic Records of the Soil Conservation Service, #149; Copyright American Economic Association, reproduced with permission of the American Economic Review from Figure 2, Hornbeck, 2012.

the central and southern plains, the classically defined Dust Bowl region, with significant areas of damage extending into Nebraska, Iowa, the Dakotas, and Montana. Much of this erosion took the form of wind erosion, which created the canonical dust storms associated with the drought. The remarkable scale and magnitude of these events can be seen in some of the reports of visible dust days across the United States (Lee & Gill, 2015). The largest dust storms, with visibility reduced to less than 1 mile, were documented by the

Soil Conservation Service (SCS) for nearly all years of the drought (Worster, 2004). The SCS recorded 14 such dust storms in 1932, 38 in 1933, 22 in 1934, 40 in 1935, 68 in 1936, 72 in 1937, 61 in 1938, 30 in 1939, 17 in 1940, and 17 in 1941 (Worster, 2004). Four major dust storms in April and May of 1934 spread dust across the eastern United States as far as North Carolina, Florida, and Washington, DC (B. Cook, Seager, & Smerdon, 2014; Mattice, 1935). Some events were so large they transported dust as far as Greenland, where deposition of the dust can be seen in ice cores (Donarummo et al., 2003).

Surveys by the SCS in 1934 found that 65 percent of the central plains was damaged by wind erosion, with 15 percent severely damaged (Hansen & Libecap, 2004). By 1938, updated estimates from the SCS indicated that 80 percent of the southern plains was affected by wind erosion, with 40 percent severely impacted (Hansen & Libecap, 2004). Soil losses may have reached as much as 850 million tons from 4.34 million acres in 1935 alone (Hansen & Libecap, 2004). Across the entire drought area, 10 million acres are believed to have lost 5 inches of topsoil, with another 13.5 million acres losing 2.5 inches, for an average loss of 480 tons per acre (Hansen & Libecap, 2004). The erosion and degradation had serious impacts on agricultural and economic productivity and fertility in the region. Transported soils generally had significantly higher levels of nutrients (e.g., nitrogen and phosphorus) and organic matter than the soils left behind, ultimately decreasing soil fertility (Hansen & Libecap, 2004; Worster, 2004). In 1933 alone, the loss of fertility associated with erosion is estimated to have reduced annual agricultural productivity by 15–25 percent, part of an estimated $400 million in productivity lost annually to erosion during the drought (Hansen & Libecap, 2004).

LAND DEGRADATION IMPACTS ON THE DUST BOWL DROUGHT

The hypothesis that land degradation itself may have contributed to the drought (i.e., by influencing precipitation and temperature anomalies) originated in the atypical pattern of drought anomalies observed during the Dust Bowl. Droughts in western North America associated with a cold tropical Pacific (as occurred during the Dust Bowl) are typically much more centered in the southwestern region (instead of the entire central plains), with major precipitation deficits occurring in the winter and spring (rather than the summer and fall). Warm Atlantic conditions are associated with summer and fall precipitation deficits, but for the 1930s, these appear more extreme than would be expected given the SST state. And even though SST-forced climate models do produce the broad features of the Dust Bowl (Schubert et al., 2004), many important details are missed in these simulations, indicating that some important processes may be missing. Also, whereas internal atmospheric variability appears to have played some role, these processes are still unlikely to explain all of the remaining features of the drought.

Land degradation influences land-atmosphere interactions in several ways relevant for drought. Greater exposure of bare soil as vegetation cover is reduced increases surface albedo, decreasing net radiation at the surface and in the atmospheric column and reducing the energy available for evapotranspiration and convection. The loss of deeply rooted vegetation can also reduce evapotranspiration, as access to deep soil moisture pools is lost. This shifts the surface energy balance to favor sensible heating, reducing moisture fluxes into the boundary layer and atmosphere and cutting off the supply of moisture needed to trigger precipitation events. Finally, the dust storms themselves may play a role, with dust particles in the atmosphere increasing the albedo through increased reflection of solar insolation. As with the albedo increases associated with reductions in vegetation cover, this can reduce net energy availability, convection, and precipitation.

Indeed, experiments with the Goddard Institute for Space Studies (GISS) climate model demonstrate that these land degradation processes have a significant impact on simulated precipitation and temperature anomalies for the Dust Bowl (figure 6.5). When forced using only the observed SST patterns ("SST Only"), the GISS model produces a drought with some modest drying over the central plains but with the most intense warming and drying centered over the Southwest. This result is qualitatively similar to the independent study of Schubert et al. (2004) in that the model fails to capture some notable features of the drought. When the model, in addition to the SST forcing, incorporates land degradation in the form of reduced vegetation

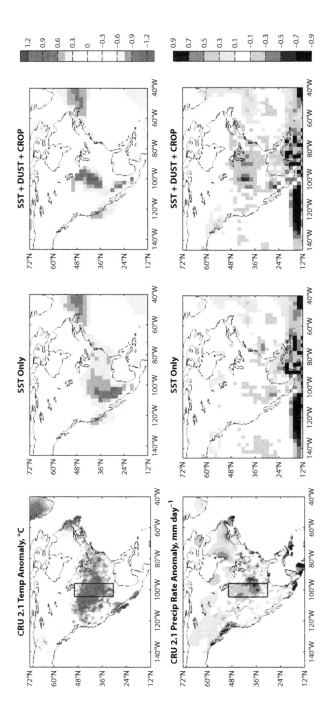

FIGURE 6.5 Temperature and precipitation anomalies during the Dust Bowl (1932–1939) from observations and climate model simulations using the Goddard Institute for Space Studies (GISS) climate model. Simulations forced with observed sea surface temperatures alone ("SST Only") produce some features of the drought but center the temperature and precipitation anomalies too far to the southwest compared to observations. When land degradation factors (dust aerosols from wind erosion and loss of vegetation cover from crop failures) are included in the simulations ("SST + DUST + CROP"), the model intensifies the drying and warming over the central plains, better matching the observations. *Source:* Reproduced from Figure 1, B. Cook et al., 2009.

cover (representing crop failure) and new dust sources (to simulate the effect of the dust aerosols) ("SST + DUST + CROP"), the drought is simulated with much higher fidelity. Precipitation deficits intensify over the central plains, covering a much broader area of the central United States in a pattern much more consistent with observations. Heating also intensifies, with much warmer temperatures over the central plains in the land degradation scenario compared to the SST-forced scenario.

After combining the GISS model results with previous analyses (Schubert et al., 2004), a likely trajectory for how the Dust Bowl began, intensified, and spread begins to emerge. The drought itself likely started as a consequence of anomalous SST forcing in the tropical Pacific (cold) and Atlantic (warm), causing modest growing-season moisture deficits across the western United States. Farmers at the time had little experience managing drought, leading to the widespread crop failures that would be responsible for large-scale land degradation and historically unprecedented levels of wind erosion and dust storm activity. The reduced vegetation cover and dust aerosols associated with the land degradation further suppressed precipitation and amplified the drought, spreading the drying across the central plains. Additional contributions from internal atmospheric variability also likely intensified the drought in some regions, including over the Pacific Coast and in the northern plains. The combination of both natural climate variability and local human activities thus turned what may have normally been a modest drought into a catastrophic event with a relatively unique seasonal and spatial signature.

RECOVERY AND MITIGATION

The end of the Dust Bowl drought and the associated recovery of ecosystems and agriculture can be attributed to three things (Lee & Gill, 2015). The first is a strong and persistent El Niño event in 1940 and 1941 that brought significant precipitation to the western United States, effectively ending the moisture deficits. The second is the economic recovery, facilitated by the 1930s New Deal reforms pioneered by President Franklin Roosevelt and the economic stimulus and rise in crop prices instigated by World War II. Finally, better land management and erosion control practices were employed that helped the landscape recover, reduced erosion, and increased agricultural resiliency in the western United States. Such reforms have been so successful that despite the significant economic and agricultural impacts of later droughts in the 1950s and at the turn of the 21st century, the central plains has not again experienced anywhere near the same level of erosion and land degradation (Hansen & Libecap, 2004).

The improvements in land management were largely pioneered by the SCS, led by Hugh Hammond Bennett, who would be later known as the "Father of Soil Conservation" (Helms, 2009). The SCS was initially founded in 1933 as the Soil Erosion Service (SES) in the Department of the Interior, later reorganized into the SCS in 1935 under the Department of Agriculture, and finally renamed the Natural Resources Conservation Service (NRCS) in 1994. In addition to organizing the first large-scale surveys and estimates of erosion damage associated with the drought, the SCS initiated a variety of activities to improve land quality and reduce erosion. On private farms, the SCS promoted erosion control measures and taught farmers best practices, including contour plowing and terracing (which increase moisture retention) and strip cropping (which reduces wind speed at the surface) (Lee & Gill, 2015). Another approach, initiated by the Prairie States Forestry Project in 1934 and expanded in 1936, was to plant rows of trees to create windbreaks or shelterbelts, reducing wind speed at the surface and wind erosion on downwind agricultural fields (Dahl, 1940; Rodgers, 2001). Where installed, these shelterbelts were effective, but their overall impact was limited because much of the central plains is simply too dry to support unirrigated rows of trees. Starting in 1935, the federal government began paying farmers to till fields for erosion control (Lee & Gill, 2015). It also urged states to establish conservation districts, communities of independently owned farms that coordinated erosion control activities (Hansen & Libecap, 2004). Kansas and Texas even allowed soil conservation districts to conduct emergency tillage operations on private land that was damaging neighboring property (Hansen & Libecap, 2004). In addition, the SCS also engaged in actively restoring grass cover through the purchase of 235,000 ha of land (a mix of croplands and rangelands) deemed unsuitable for continued or future farming (Lee & Gill, 2015). These purchases occurred in the late 1930s and early 1940s, and, once revegetated, the land was leased to ranchers for grazing (Lee & Gill, 2015).

Other advancements, initiated during the 1930s or soon thereafter, made the region more resistant to erosion during future droughts. New plow designs left more crop residue at the surface (increasing surface roughness and reducing wind speeds) and brought more clays to the surface (increasing soil cohesion and resistance to the wind) (Lee & Gill, 2015). The widespread practice of irrigation, often using groundwater from deep aquifers, improved crop productivity and erosion control during droughts (Lee & Gill, 2015). Finally, various government programs incentivized the conversion of cultivated land to perennial vegetation and ecosystems. These include the program to establish National Grasslands, as well as other set-aside programs (e.g., the Soil Bank Program from the 1950s to the 1970s and the Conservation Reserve Program from 1985 to the present) through which the government pays farmers to plant or maintain native or noncultivated vegetation on part of their land (Lee & Gill, 2015).

The Sahel

The Sahel is a semiarid region of West Africa, typically defined as a narrow expanse of land between 10°N and 20°N latitude and extending from Mauritania and Senegal on the Atlantic Coast to 10–30°E longitude (figure 6.6) (Giannini, 2016; Nicholson, 2013). Climatically, the Sahel represents the most northward extent of the West African monsoon climate; it comprises a transitional zone of semiarid grassland, shrubs, and savanna vegetation bordered by moist tropical forests to the south and the much drier Sahara Desert to the north. Within the Sahel, there is a strong meridional gradient in precipitation, with ~600 to 1,000 mm/year at the southern limit and 100 to 200 mm/year at the northern edge, in sharp contrast to the strong east-west uniformity in both climate and vegetation. Historically, the Sahel has been home mostly to subsistence farmers and pastoralists, populations that are typically highly vulnerable to variations in climate and weather. Because it is affected by the West African monsoon, precipitation in the Sahel is

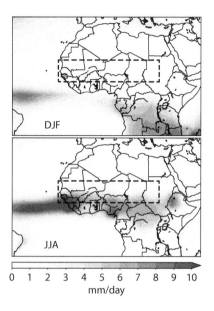

FIGURE 6.6 Maps of seasonal average precipitation (1979–2017) for boreal winter (December–January–February; DJF) and boreal summer (June–July–August; JJA), centered over West Africa. The Sahel region of West Africa (indicated by the dashed black box; 10°N–20°N, 18°W–30°E) lies in the transition zone between the relatively wet sub-Saharan Africa to the south and the much more arid Sahara Desert to the north. Climate in the Sahel is strongly affected by variations in the West African monsoon, and this region is home to mostly populations of subsistence farmers and pastoralists. *Source*: Global Precipitation Climatology Project, National Center for Environmental Information, https://www.ncei.noaa.gov/data/global-precipitation-climatology-project-gpcp-monthly/access/.

highly seasonal. During the boreal winter, the Intertropical Convergence Zone (ITCZ) is shifted toward the equator and south of the Sahel, resulting in an extended dry season from November through March. The wet season occurs during boreal summer when the ITCZ shifts farther north and southwesterly winds bring moisture into the region. Most precipitation in the Sahel falls in July, August, and September, and year-to-year variability in total precipitation is most closely connected to variability in precipitation during these months (Nicholson, 2013). Vegetation productivity and the growing season are tightly coupled to the seasonal supply of moisture in the monsoon, with peak productivity occurring typically toward the end of the monsoon season.

THE LATE 20TH-CENTURY SAHEL DROUGHT

The onset of the Sahel drought in the late 1960s represents one of the most abrupt climate shifts documented anywhere in the historical record (figure 6.7). The 1950s and most of the 1960s were a period of widespread and spatially coherent anomalous wet conditions across the Sahel, with above-average precipitation in nearly every year from 1950 to 1967 (Nicholson, 2013). The drought itself began in 1968, and precipitation deficits in the region persisted more or less continuously until 1997 (Nicholson, 2013). Precipitation during much of the 1980s was only ~60 percent of average, whereas over the longer term (1968–1997), deficits in August precipitation were (depending on the region) from 37 to 55 percent below the average from previous decades (1931–1960) (Nicholson et al., 2000). The overall drying corresponded to a southward shift of 1 to 2 degrees of latitude in rainfall isohyets (Nicholson, 2013). Despite this drought's extreme nature, however, historical records suggest that similarly persistent droughts may have occurred in the Sahel in the early 1800s (Nicholson et al., 2012).

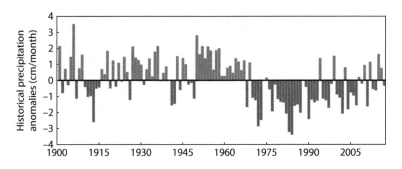

FIGURE 6.7 Monsoon season (June–October; JJASO) historical precipitation anomalies (relative to the 1901–2017 mean) averaged across the Sahel (10°N–20°N, 20°W–10°E) from 1901 to 2017. Anomalously wet conditions affected the Sahel for most of the 1950s and 1960s, but the region experienced an abrupt shift toward drought in 1968. Despite some recovery from the peak precipitation deficits in the 1970s and 1980s, this late 20th-century drought persisted well into 1990s, resulting in a mostly continuous ~30-year period of low precipitation. Source: Data from the University of Washington Joint Institute for Study of the Atmosphere and Oceans, doi:10.6069/H5MW2F2Q, http://research.jisao.washington .edu/data_sets/sahel/. Underlying data are from the Global Precipitation Climatology Centre Full Data Reanalysis Version 7 [Schneider et al., 2015] and First Guess Product [Ziese et al., 2011], https://www .dwd.de/EN/ourservices/gpcc/gpcc.html.

IMPACTS OF THE DROUGHT

The Sahel drought had significant impacts on ecosystem health and food security, creating humanitarian crises across West Africa that motivated much of the initial interest in the event from the international and scientific communities. The effects of the drought across the Sahel, however, were not uniform and were strongly modulated by a variety of nonclimatic (social, political, and economic) factors that affected local resilience (Gautier et al., 2016). In many areas, communities responded to the drought by diversifying or switching their primary activities (e.g., changing from agriculture to wood harvesting), migrating (either temporarily or permanently), and (where possible) taking advantage of support and aid provided by local, national, and international governments and organizations (Gautier et al., 2016).

The drought had a pronounced negative effect on ecosystems. Broadly, biodiversity (species richness) in the Sahel declined from 1945 to 2002 (Epule et al., 2014). In Guinea, diversity of mesic trees and shrubs declined by ~54 percent, and, in Sudan, tree diversity declined by ~29 percent. In Senegal, tree density declined by 29 percent between 1972 and 1976 and by 20 percent between 1954 and 2002 (Gonzalez et al., 2001 Wezel & Lykke 2006). The drought caused a per capita food production decline of ~25 percent across West Africa in the 1970s and 1980s (Dilley et al., 2005; Epule et al., 2014). Nigeria experienced a 60 percent decrease in crop production during the 1970s (Obioha, 2009; Tambo &. Abdoulaye, 2013). In 1984 alone, millet production declined by 70 percent in Burkina Faso (Swinton, 1988; Webb & Reardon, 1992), rice yields collapsed in Senegal (Kasei et al., 2010), and groundnut production decreased by 50 percent in Gambia (Aubee & Hussein, 2002). Crop production across the Sahel during 1991–1992 decreased by 54 percent (Afifi, 2011). The drought also had severe impacts on pastoralists and domesticated grazing animals. During the 1970s, 32 percent of cattle and 37 percent of goats across the Sahel died, with cattle mortality as high as 80 percent in some locales, causing 2 million herders to lose their livelihoods (Bernus, 1990; Jean, 1985; Jouve, 1991; Juul, 1996; Sheets & Morris, 1974). In 1983 and 1984, 75 percent of the Sahelian herds either died or were sold (De Bruijn, 1997). Because of the declines in food production, 22 million people suffered from famine during the 1970s (Aubee & Hussein, 2002; Reardon et al., 1988), with another 17–19 million people at risk for starvation during 1990–1992 (Afifi, 2011; Webb & Reardon, 1992). In all, an estimated 100,000 deaths have been attributed to the Sahel drought during the 1970s and 1980s (Afifi, 2011).

The environmental and humanitarian crisis created by the Sahel drought inspired both regional and global responses. Much of the motivation came from the generally held belief at the time that the desertification associated with the drought was anthropogenic in origin (e.g., caused by overgrazing) and that this degradation acted as a feedback to amplify the drying (Giannini et al., 2008; see the next section). Early in the drought (1973), nine countries in the Sahel region (Burkina Faso, Cape Verde, Chad, Gambia, Guinea-Bissau, Mali, Mauritania, Niger, and Senegal) formed the Comité Permanent Inter-États de Lutte contre la Sécheresse dans le Sahel (CILSS; Permanent Interstate Committee for Drought Control in the Sahel) (Giannini et al., 2008). The primary mandate of the CILSS is to invest in ways to improve food security in the region and increase agricultural and ecological resilience to drought and desertification (Giannini et al., 2008). The United Nations (UN) also took a keen interest in the drought, convening the first UN Conference on Desertification in Nairobi in 1977. This would eventually result in the establishment of the UN Convention to Combat Desertification (UNCCD) in 1994, with the specified goal "to forge a global partnership to reverse and prevent desertification/land degradation and to mitigate the effects of drought in affected areas in order to support poverty reduction and environmental sustainability."

CAUSES OF THE SAHEL DROUGHT

From the beginning, desertification was considered a likely cause of the intensity and persistence of the Sahel drought. At the time, the land degradation was primarily attributed to direct human activities, including agricultural expansion, overgrazing, and wood harvesting (Giannini et al., 2008). These activities would have reduced vegetation cover, affecting the surface energy balance, and land-atmosphere interactions and potentially precipitation. This hypothesis was first formalized by Jule Charney (1975), where he proposed that a decrease in plant cover would increase surface albedo, reducing net incoming radiation and increase radiative cooling of the air, causing subsidence that would suppress convection and precipitation.

To test these ideas, Charney conducted two experiments using a climate model in which the surface albedo over the Sahel region was modified to mimic the effects of desertification. In the "control" scenario, albedo over the Sahel was set to 0.14; in the "perturbation" scenario (representing desertification), the albedo over the same region was increased to 0.35. As hypothesized, precipitation was significantly reduced over the Sahel in the high-albedo perturbation experiment, ostensibly supporting the desertification-drought hypothesis. Results from this simplified experiment, however, were amended in later decades as our understanding of climate and drought in the region improved and our modeling capabilities advanced. For example, Govaerts and Lattanzio (2008) compared albedo between one of the worst years of the drought (1984) and an anomalously wet year (2003) and found only a 0.06 difference. This suggested that the albedo perturbation in the Charney (1975) experiment was too large and likely overestimated the impact of desertification on the drought. Later, new model experiments using more sophisticated treatments of the land surface and vegetation (e.g., Giannini et al., 2003; G. Wang & Eltahir, 2000; Zeng et al., 1999) attributed the desertification amplification of the drought to processes other than albedo, including feedbacks from vegetation dynamics and soil moisture.

More recently, a new consensus has emerged that relegates the desertification hypothesis to a secondary, amplifying role and highlights SST anomalies in the tropical oceans as the primary cause of the Sahel drought (Giannini et al., 2008). Drought years in the Sahel have long been associated with warm SSTs in the equatorial Atlantic (Hastenrath & Lamb, 1977; Lamb, 1978a, 1978b). Because SST anomalies in the Atlantic Ocean evolve slowly, reflected in indices like the Atlantic Multidecadal Oscillation (AMO), Atlantic SSTs have an especially strong influence on Sahel rainfall on decadal and longer timescales (Giannini, 2016). For the Sahel, the relevant pattern over the Atlantic is the meridional gradient in SSTs between the northern and southern tropical oceans because this gradient influences the location of the ITCZ. When the tropical South Atlantic is warmer than the tropical North Atlantic, the ITCZ is preferentially displaced southward, and the monsoon over West Africa and the Sahel is weaker, favoring drought. The Indian Ocean also affects the Sahel—when the equatorial Indian Ocean is warmer than normal, strong convection occurs over the oceans in the tropics. Warming occurs throughout the tropical troposphere, increasing vertical stability and the threshold for convection. If this threshold cannot be met by air parcels over land (which is often the case because the land is drier than the oceans), the result will be a decrease in terrestrial precipitation and an increase in the likelihood of drought (Giannini, 2016).

During the 1970s and 1980s, conditions in the Atlantic Ocean (cold tropical North Atlantic and warm tropical South Atlantic) and the Indian Ocean (warm) were ideally configured to cause drought over the Sahel and West Africa. The connection between Atlantic SSTs and the Sahel drought is further supported by the close multidecadal synchronicity between precipitation variability over the Sahel and the AMO index, with drought in the early 1800s and early 1900s coinciding with extended negative AMO conditions (indicative of cool conditions in the tropical North Atlantic). When climate models were forced with the observational estimates of SSTs at the time (e.g., Giannini, 2003), they were able to reproduce the general persistence and magnitude of the Sahel drought, along with other periods of persistent hydroclimate variability in the region during the 20th century (e.g., the wet 1950s and 1960s in the Sahel and the recovery from the drought in the late 1990s).

The convergence of evidence indicates that the late 20th-century Sahel drought was forced primarily by patterns of SST variability but that land surface feedbacks from desertification acted as a secondary drought amplifier. A remaining question to resolve is the extent to which anthropogenic climate change may have contributed to the drought (e.g., through the influence of greenhouse gas warming on SST trends). The warming in the Indian Ocean associated with the Sahel drought is mostly attributed to greenhouse gas–driven increases in ocean heat content (Barnett et al., 2005). Attributing warming trends over the Atlantic (e.g., in the South Atlantic) to climate change is more difficult because of strong natural variability (e.g., the AMO), which is difficult to separate from the forced warming signal. There is evidence, however, that the cool anomaly in the North Atlantic during the Sahel drought, which would have contributed to the southward displacement of the ITCZ, was caused, at least in part, by anthropogenic sulfate aerosol emissions (Booth et al., 2012; Chang et al., 2011). The 1970s and 1980s were a period of relatively high sulfate aerosol concentrations over the North Atlantic, which would have increased the reflection of solar radiation and cooled the surface as a first-order effect. This cooling effect began to diminish in the 1980s with the reduction in aerosol concentrations associated with legislation to reduce air pollution in the United States and Europe (Booth et al., 2012; Chang et al., 2011).

Climate models have reproduced the Sahel drought when these forcings have been applied. Biasutti and Giannini (2006) examined a suite of coupled ocean-atmosphere models that were allowed to internally generate their own SST history but that included anthropogenic sulfate and greenhouse gas emissions. Twelve of these models produced drying over the Sahel consistent with the observed late 20th-century drought, thanks in part to the impact of these forcings on the model SST evolution. These results were confirmed in the next generation of Coupled Model Intercomparison Project version 5 (CMIP5) models, although the multimodel mean in this ensemble under-estimates the magnitude of both the 1950s pluvial and the 1970s–1980s drought (Biasutti, 2013). The two clearest anthropogenic climate change signals in the Sahel drought therefore appear to be a global warming signal, manifesting in Indian Ocean warming, and a more regional sulfate aerosol signal, implicated in the North Atlantic cooling.

GREENING OF THE SAHEL

Over the past several decades, the Sahel has experienced a pronounced recovery in vegetation health and productivity (greening) (Giannini et al., 2008; Herrmann et al., 2005; Olsson et al., 2005; Polgreen, 2007). This phenomenon can be most clearly seen in vegetation indices from multidecadal satellite observations, such as the Normalized Difference Vegetation Index (NDVI) (figure 6.8). Areas of statistically significant increases in NDVI from 1981 to 2011 (indicative of increased vegetation cover or productivity) can be see across the Sahel. Significant declines are much more localized and, for the Sahel, are isolated primarily in western Niger and central and eastern Sudan (Dardel et al., 2014).

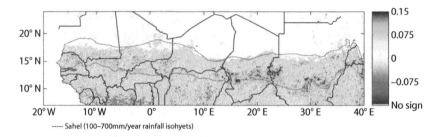

FIGURE 6.8 Linear long-term trends (1981–2011) in vegetation productivity from satellite estimates, as indicated by the Normalized Difference Vegetation Index (NDVI). Most of the Sahel has experienced significant greening and increased productivity over the last ~30 years (green regions), a result of the recovery from the drought and other factors, including land-use change. *Source:* Reprinted from Figure 7, Dardel et al., 2014.

The causes of these increases, and the processes on the ground represented by the satellite-based estimates of greening, are difficult to determine. Broadly, the greening is congruent with a positive trend in precipitation since the 1980s, most likely representing a recovery from the most extreme period of the drought in the 1980s (Giannini et al., 2008). This appears to be the case for the Gourma region of Mali, where greening is associated with increased precipitation and where the satellite trends largely reflect increases in herbaceous productivity during the peak of the growing season (August–October) (Dardel et al., 2014). Over Senegal, however, much of the greening appears to be driven by a doubling of leafy biomass in woody species, with herbaceous biomass showing significant year-to-year variability but no clear trend (Brandt et al., 2015).

Greening in some regions may be related to factors other than climate, like land-use change. For example, active reforestation efforts in some countries have led to the expansion of trees suited to arid situations (Gonzalez et al., 2001; Hiernaux et al., 2009; Wezel & Lykke, 2006). In other regions, long-term trends in vegetation do not show monotonic increases or declines. For example, in the Fakara region of Niger, significant greening was observed in the years immediately following the Sahel drought, but significant declines have been observed since 1990, despite continued increases in precipitation (Dardel et al., 2014). This decline is possibly related to other changes at the surface, including shifts in land use and management, increased grazing pressure, or overall declines in fertility (Dardel et al., 2014).

THE FUTURE OF SAHEL RAINFALL

Analyses of previous generations of climate models noted significant divergence and uncertainty across models for climate change projections of Sahel precipitation (e.g., Biasutti & Giannini, 2006; K. Cook & Vizy, 2006). Although most of these models performed reasonably well in reproducing the observed late 20th-century drying, some models predicted persistent dryness for the end of the 21st century, whereas others indicated increased precipitation (Giannini et al., 2008). The reasons for this divergence are not completely clear, but there is some evidence that it may be due to differences in the warming of the tropical oceans relative to the global mean across the models comprising the ensemble (Biasutti & Giannini, 2006; Giannini et al., 2008). In the most recent generation of models (CMIP5), however, a much stronger convergence has emerged (Biasutti, 2013). Under the high-emissions, high-warming RCP 8.5 scenario, nearly all models show declines in precipitation in early monsoon season (June–July) and increases in precipitation during the latter part of the monsoon (September–October). The early season drying is comparable in magnitude to the declines associated with the late 20th-century drought and is focused primarily in the western Sahel, whereas significant wetting occurs across the entirety of the Sahel (Biasutti, 2013). In aggregate, these models indicate an effective shift in the timing of the West African monsoon, with a delayed start but an overall wetter peak in precipitation. The ramifications of this shift in timing for ecosystems and agriculture in the region, however, are unclear.

SEVEN

Land Degradation and Desertification

L
and degradation is defined as a sustained, sometimes irreversible, loss of ecological productivity and ecosystem services. It can be caused by shifts in soil properties (e.g., erosion and soil compaction), vegetation (e.g., replacement of grasslands by shrubs), and/or climate (e.g., drought) and is often directly related to human activities (e.g., agriculture and overgrazing). Drylands (arid, semiarid, and dry subhumid areas) are especially susceptible because of their marginal productivity and limited resource availability; land degradation in these regions is defined as desertification by the United Nations Convention to Combat Desertification (UNCCD). Given the strong connection among hydroclimate, drought, and land degradation, here we discuss the drivers, feedbacks, and impacts of land degradation. What controls the susceptibility of different regions to degradation, and what are the dominant natural and anthropogenic causes? How is land degradation assessed? What feedbacks in the Earth system can amplify degradation? How will climate change affect land degradation trends in the future? And how can land degradation be ameliorated or even reversed?

Overview of Land Degradation and Desertification

Land degradation manifests in a variety of ways, but all cases involve some fundamental *reduction or loss of ecosystem services* (D'Odorico et al., 2013) that (1) provide food and water, (2) regulate climate and nutrient cycles, (3) support human health and agriculture (e.g., inhibit pest populations and support pollinators for crops), or (4) provide nonphysical benefits, such as support for spiritual or recreational activities. Degradation can take the form of soil losses from wind and water erosion, reductions in soil quality (e.g., loss of nutrients and organic matter, reductions in water-holding capacity, and increases in salinity and toxicity), and changes in vegetation composition and coverage (e.g., increased bare soil area and shifts in species composition) (D'Odorico et al., 2013). Losses from degradation can act to reinforce the degraded state through local (and sometimes regional) feedback processes that inhibit the ability of systems to recover to previous, nondegraded states. Such feedbacks may be strong enough to prevent recovery even when the initial cause of the degradation (e.g., overgrazing) is removed from the landscape. Land degradation and its impacts thus encompass an array of complex phenomena involving the interplay of many biological and physical factors over a range of spatial and temporal scales.

An illustrative recent example of land degradation is that of the Chihuahuan Desert region of southwestern North America (Ravi et al., 2010). Here, the degradation has been driven by *shrub encroachment*, a phenomenon whereby the native grasslands are progressively replaced by shrubs. The causes of shrub encroachment are not entirely understood but may be related to disturbance from overgrazing of livestock or changes in fire regimes (Ravi et al., 2010). As shrubs replace grasses as the dominant vegetation type, the landscape becomes increasingly fragmented, with less continuous vegetation cover and increased exposure of bare soil. Soil resources (moisture and nutrients) become concentrated around the vegetated patches, creating "islands of fertility" (Schlesinger et al., 1990) even as wind and water erosion rates increase significantly over the bare soil areas (Ravi et al., 2010). The end result is a landscape that is often less productive, less connected, less fertile, less biodiverse, and less suitable for supporting many ecological and human activities.

Drylands are especially vulnerable to land degradation (figure 7.1) because such regions typically have few water resources, low vegetation productivity, and high climate variability (e.g., recurring droughts). Ecosystems in these regions are thus highly susceptible to the types of disturbances that can cause desertification. Areas of particular risk include large swathes of western North America, sub-Saharan Africa, the Middle East, South America, and Australia. Despite the limited extent and relatively low productivity of degraded drylands, their impact on human populations is disproportionately large. This is because ~35 percent of the global population (~2 billion people) lives in drylands and depends on these ecosystems for their livelihoods and another ~20 percent of the global population lives in areas that have experienced some degree of desertification (D'Odorico et al., 2013).

Large-Scale Detection and Monitoring of Degraded Lands

At the surface, detection and monitoring of land degradation is based on biophysical and social indicators that demonstrate losses of productivity or ecosystem services (D'Odorico et al., 2013). Biophysical indicators include vegetation and land-cover changes (e.g., increased bare soil area), losses of biodiversity, and reductions in soil fertility. Examples of economic and social factors are declining crop yields, losses of revenue and jobs, emigration, and declines in health. Data can come from surveys and field measurements, interviews, and remote sensing (D'Odorico et al., 2013). Such detailed monitoring, however, depends on having the resources (e.g., money and labor), access, and expertise necessary to conduct fine-scale on-the-ground assessments, requirements that are typically prohibitive for assessing land degradation at much larger spatial scales.

At the regional and global level, the challenge is to connect broad, large-scale datasets of environmental change to fundamental degradation processes operating on the ground. This connection is not always clear, however, because (as mentioned previously) degradation can manifest in a variety of ways. For example, satellite datasets of vegetation activity, like the Normalized Difference Vegetation Index (NDVI), can provide important information on changes in vegetation productivity and cover. Declining

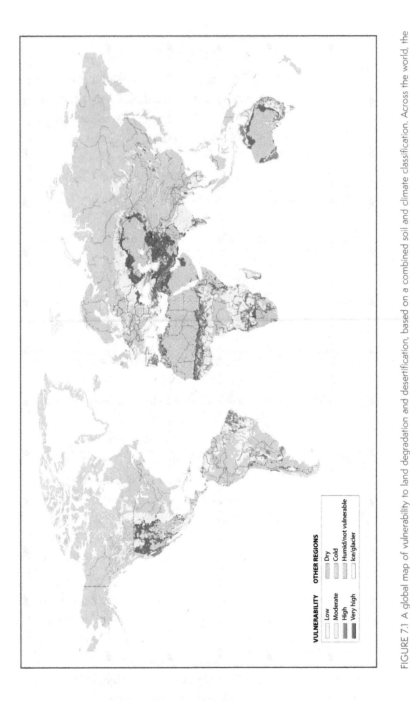

FIGURE 7.1 A global map of vulnerability to land degradation and desertification, based on a combined soil and climate classification. Across the world, the most vulnerable regions are in semiarid and arid locations, highlighting the importance of hydroclimate. These include areas of western North America, Australia, the Middle East, and sub-Saharan Africa. *Source:* USDA-NRCS, Soil Science Division, World Soil Resources, Washington, DC, https://www.nrcs.usda.gov/wps/portal/nrcs/detail/soils/use/worldsoils/?cid=nrcs142p2_054003.

trends or shifts in NDVI in a region can therefore be used as an indicator that land degradation is occurring. However, these NDVI trends can be difficult to connect to other important degradation processes (e.g., changes in soil fertility), often cannot be used to distinguish changes in vegetation composition (e.g., shifts in species or plant type), and must be combined with other datasets to identify causal factors driving the observed changes. Alternative methodologies typically deployed for global assessments (e.g., expert judgment and surveys of abandoned agricultural lands) have their own limitations. There is thus no definitive approach for assessing global land degradation trends and no consensus on the most accurate methodology. Large-scale assessments of land degradation generally rely on four different approaches with varying strengths and weaknesses: expert opinion, satellite observations, biophysical models, and inventories of abandoned agricultural lands (Gibbs & Salmon, 2015).

Expert opinion relies on the often-subjective judgments provided by local experts on various characteristics (e.g., type, extent, degree, and cause) of degradation, which are then integrated to generate global and regional estimates. The Global Assessment of Soil Degradation (GLASOD), commissioned by the United Nations Environment Program (Oldeman, 1992; Oldeman et al., 1990), was one of the first large-scale attempts to use expert judgment. According to GLASOD, hot spots of degradation are centered in Central America, sub-Saharan Africa, the Middle East, and Asia. Because of the subjectivity inherent in the expert assessments underlying GLASOD, however, separate land degradation estimates based on this dataset often disagree widely. Oldeman and colleagues (1990), for example, estimated from GLASOD that ~2 billion ha of land have been degraded since the middle of the 20th century, representing ~22.5 percent of global agricultural land, pasture, and forest and woodland. A later assessment (Oldeman & van Lyden, 1997) revised this estimate down to 1.14 billion ha. An alternative analysis of GLASOD (Bot et al., 2000), however, concluded that over 6 billion ha have been affected by degradation, about three times the original estimate of Oldeman et al. (1990).

The application of satellite observations in tracking land degradation generally relies on indices of vegetation activity, such as NDVI. Satellites have the advantage of enabling temporally and spatially continuous estimates of land degradation in a consistent and quantitative manner. This is especially valuable for investigating degradation in many isolated and developing regions, where access to high-quality observations and assessments on the ground may be limited. An example of this approach is the United Nations Global Assessment of Land Degradation and Improvement project (GLADA) (Bai et al., 2008). In GLADA, land degradation trends are defined as long-term declines in ecosystem function, using satellite NDVI as a proxy for net primary productivity. These trends are estimated from residual changes in NDVI after climate variability and land-use shifts are removed; the focus is therefore on degradation from direct human (nonclimatic) impacts (e.g., overgrazing). According to GLADA, an estimated 2.7 billion ha have experienced degradation, a figure similar to some of the independent estimates based on expert opinion using the GLASOD dataset. However, the spatial distribution of the major centers of degradation in GLADA is different, with the most intense regions of degradation occurring in more humid areas (e.g. equatorial Africa and China). Satellite-based estimates of degradation have some significant limitations. The satellite record itself is relatively short, beginning in the late 1970s and early 1980s, and thus cannot provide information on degradation that occurred prior to these dates. The satellite data themselves suffer from a variety of issues that must be accounted for to develop a continuous and robust record of vegetation activity, including the replacement of satellites and instruments and the contamination and interference from atmospheric effects (e.g., clouds). Degradation estimates from satellites can further be confounded by issues on the ground that may be difficult to parse or detect from the satellite observations alone. These latter issues are likely to vary from region to region and can include shifts in land use unrelated to degradation, agricultural intensification, and declines in soil fertility.

Biophysical modeling uses climate and soils data to estimate land degradation by predicting vegetation and agricultural productivity. Using such an approach, Cai and colleagues (2011) identified ~1 billion ha globally as potentially degraded. However, the spatial distribution differs from those of GLASOD and GLADA, with the most intense degradation in India, extratropical South America, Europe and eastern Russia, and China. Compared to expert opinion and satellite estimates, biophysical modeling is a relatively simple approach using independent climate and physical data rather than any direct observations of land degradation. As such, its validity can depend critically on the quality of data used, which is often poor in the data-sparse regions of developing countries where land degradation is a major concern. Further, because biophysical modeling relies on indirect data, it can provide only estimates of *potential* degradation. Biophysical modeling therefore must be ground validated or reconciled with other independent methods (e.g., GLADA) to increase confidence in these estimates.

Finally, land degradation can be assessed by tracking the abandonment of agricultural lands, based on the implicit assumption that such lands were abandoned because of declines in productivity and fertility. This approach usually relies on the combined use of satellite data on vegetation changes and agricultural census data. Estimates using this approach (Campbell et al., 2008) suggest that 269 million ha of cropland and 479 million ha of pastureland were abandoned over the last 300 years, a result lower than those from the other three methods. Degradation estimates based on agricultural land abandonment, however, are likely to be conservative and underestimate total degradation because this method excludes degraded lands that have not been abandoned. It also omits changes in land use in the preindustrial era (when data are unavailable) and excludes shifting cultivation, a process whereby land is used for several years and then allowed to go fallow to recover.

Land Degradation Processes

Although it may be difficult to robustly estimate global-scale trends in land degradation, the processes by which lands can become degraded are broadly understood. These include mechanisms (figure 7.2), both natural and anthropogenic, that can initiate degradation (e.g., drought and overgrazing) and feedbacks that can intensify an existing degraded state or make it difficult for the landscape to recover. Direct human dimensions fall into two broad categories. The first, land-cover conversion, is the complete replacement of one land-cover type by another, such as the clearing of a forest for cultivated crops or pasture. The second, land management, involves the introduction of new activities onto the landscape, such as the grazing of domestic livestock or irrigation of croplands. Feedback mechanisms operate both locally and regionally and are related to (1) shifts in soil

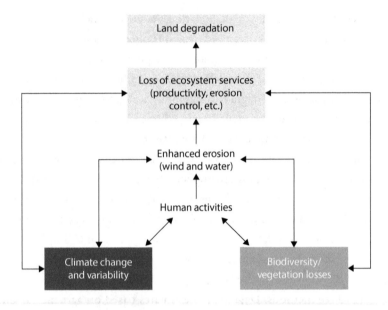

FIGURE 7.2 Interactions among humans, climate, and land surface processes that can initiate or contribute to land degradation and desertification. Human activities that can play important and direct roles in triggering land degradation include overgrazing, poor agricultural management, and salinization through irrigation water. These effects can amplify the degraded state through regional climate feedbacks and further declines in vegetation productivity, biodiversity, erosion control, and soil quality. Humans may also contribute indirectly through climate change—induced increases in aridity and drought. *Source*: Adapted and modified from Figure 2, Ravi et al., 2010.

properties, (2) coupled interactions between the land surface/vegetation and climate, and (3) changes in the biodiversity and species composition of the local plant communities.

Local Drivers

Land conversion is perhaps the most dramatic and obvious human impact on the landscape. Globally, there has been a widespread conversion of natural ecosystems (e.g., forests and grasslands) to cultivated croplands and pastures. Of these, pasture (land used for grazing domesticated livestock) is the most widespread human land use, accounting for ~25 percent of the global land area, or ~2.5 times the area cultivated for crops. Human land-use change is not a recent phenomenon, and many regions (e.g., Europe, India, China, and the Middle East) have land-use legacies extending back thousands of years (Ellis et al., 2013), with recorded instances of land degradation and soil erosion dating back almost as long (Ravi et al., 2010). Other regions (e.g., North America, Australia, and extratropical South America) have seen widespread land-use change only in more recent centuries, and several areas have actually "recovered," moving toward a more natural vegetation state. An example of the latter is the eastern United States, which was a major center of agricultural production until the early 20th century, when the majority of agriculture shifted to the central plains (discussed in more detail in chapter 6). As a result of this land abandonment, much of the eastern United States has steadily reforested over the last century. Human impacts on the landscape, and the associated land degradation, are therefore not solely a recent phenomenon and extend back centuries or even millennia in some regions.

Vegetation provides a variety of important erosion control services. Roots hold soil particles together, making them more resistant to erosion by wind and water, and create channels that increase the infiltration and water-holding capacities of the soil, reducing ponding and runoff at the surface. Organic matter also increases the total water-holding capacity of the soil, allowing for greater absorption of moisture. Aboveground, vegetation intercepts wind and water (precipitation and runoff), absorbing energy, reducing the impact of these erosion agents on the soil, and impeding sediment transport. Erosion, especially from wind, is inhibited when soils are wet because of the bonding properties of water that help hold soil particles together. Many of these services are lost or their efficacy is severely reduced when the landscape is cleared and converted to more intensively managed croplands and pastures, especially if vegetation coverage on the landscape declines and more bare soil is exposed. Globally, erosion has affected ~87 percent of degraded lands, with 1,094 million ha affected by water and 550 million ha by wind (Middleton & Thomas, 1997). One of the starkest examples of how land conversion can dramatically increase erosion rates can be found in Central America during the height of the Maya civilization (figure 7.3). The Maya expanded and intensified agriculture in the Lake Salpetén catchment in Guatemala from ~2000 BCE to the early Common Era, indicated by the presence of *Zea* maize pollen in lake sediment cores. This land-cover conversion, and subsequent disturbance to the soils, caused

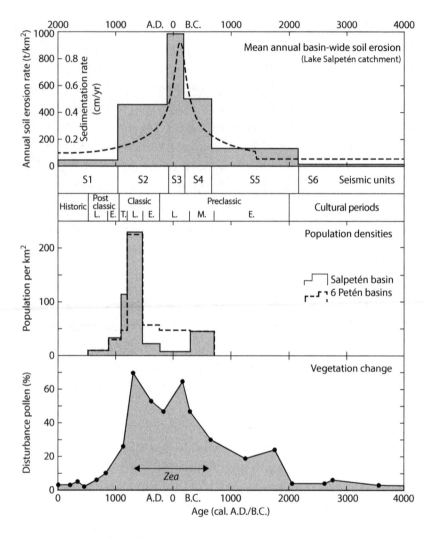

FIGURE 7.3 Estimated annual soil erosion rates, population density, and disturbance pollen (Zea spp.; maize) from the Lake Salpetén catchment in Guatemala from 4000 BCE to the present day. Widespread increases in agricultural land use (indicated by the increased abundance of Zea pollen from 2000 BCE to 700–1000 CE) caused soil erosion rates to increase to 20 to 50 times the natural rates during the transition to the Common Era. Source: Figure 3, Anselmetti et al., 2007.

significant increases in annual erosion, peaking at rates ~20 to 50 times the natural background levels (Anselmetti et al., 2007). Erosion rates declined after the Maya collapse ~1000 CE, as agricultural land was abandoned and the natural vegetation recovered.

Beyond wholesale land conversion, shifts in land management can also cause land degradation. One of the most often cited contributors to degradation is overgrazing, a broadly and somewhat vaguely defined term that refers to the stocking of animals (usually domesticated livestock) to a level beyond the support capacity of the landscape. In many natural systems, grazing animal populations are limited by the forage and water resources available, which encourages constant movement of herds in search of food and water. This constant movement allows time for vegetation to recover, limiting the intensity of herbivore impacts. However, these biophysical limitations are often ignored when stocking domesticated livestock (e.g., sheep and cattle), resulting in denser and less mobile herds and larger impacts on the vegetation and soils. The effect of overgrazing is often most severe near artificially installed watering holes, around which livestock will tend to congregate (D'Odorico et al., 2013).

Impacts of overgrazing manifest in three primary ways: defoliation, treading, and excretion (Bilotta et al., 2007). The grazing animals' consumption of and damage to aboveground vegetation (e.g., from crushing and trampling) and root systems (e.g., from soil compaction) can cause excessive defoliation, negatively affecting vegetation health, productivity, and coverage. In some cases, this can cause reductions in biodiversity and species composition (e.g., a shift from palatable to unpalatable grass species). With reduced vegetation cover, wind and water erosion increases, resulting in losses of soil resources. The soil compaction resulting from the treading of hooves can also reduce the hydraulic conductivity and water-holding capacity of the soil, making it difficult for water to infiltrate and leading to increased surface runoff, which can further contribute to erosion. Finally, excretion of animal waste can significantly alter the nutrient balance of the landscape and can cause pollution of waterways as the waste is transported (e.g., by the enhanced surface runoff). Overgrazing itself is a continuous process, with the level of degradation and impact on soils, vegetation cover, and plant community composition progressively increasing over time. One region where overgrazing appears to be significantly contributing to recent land degradation trends is Mongolia (Hilker et al., 2014). From 2002 to 2012, populations of grazing animals increased while, concurrently, vegetation productivity (as measured by satellite NDVI) in many parts of the country declined (Hilker et al., 2014).

Over both grasslands and regions with more sparse vegetation, these NDVI declines are linked strongly to the cumulative change in herd size, indicating that ~80 percent of the degradation can be attributed to increases in livestock populations and overgrazing (Hilker et al., 2014).

Another common cause of land degradation is soil salinization, which refers to the accumulation of salts in the upper soil profile and root zone. High salt content in the topsoil inhibits plant establishment and growth, negatively affects crop yields, and contaminates surface waters (D'Odorico et al., 2013). Salinization can occur in areas with shallow groundwater tables (less than 2.5 m), where capillary action brings dissolved salts upward into the topsoil and root zone. In areas with deep groundwater and poor drainage, salinization can be caused by precipitation, weathering near the surface, and dry deposition, especially when drainage is poor and evaporative demand is high, allowing salts to concentrate in the surface soil. A common cause of soil salinization that is especially relevant from a human land degradation perspective is irrigation. Many marginal and semiarid lands require water from irrigation to support cultivated crops, and any salts dissolved in the irrigation water will be left behind when this water is evapotranspired. This effect can be further exacerbated if poor-quality, untreated groundwater with especially high concentrations of solutes is used for the irrigation. Globally, salinization is estimated to affect between 45 and 831 million ha (D'Odorico et al., 2013; Sivakumar, 2007).

Coupled interactions between soils and vegetation drive many of the local positive feedbacks that can lead to the persistence of degraded states. For example, the loss of erosion control services from removal of or damage to vegetation can result in nearly irreversible losses (at least on decadal timescales) of soil fertility and soil quality (e.g., organic matter content and soil water-holding capacity). The reductions in soil quality can make it difficult for the vegetation to recover or reestablish previous states, resulting in systems that are locked into a relatively "stable" state of degradation with increased bare soil area and enhanced erosion rates. Fertility islands are a classic example of this, with isolated pockets of vegetation acting to concentrate nutrients and soil moisture, whereas the bare soil areas mean that total erosion across the landscape is increased. Similarly, salinization and loss of soil nutrients through extraction of crops can degrade soils and limit the ability of vegetation to recover or reestablish, at least in the absence of active soil remediation. At the landscape scale, losses of vegetation and soils from land degradation can change the "functional connectivity" of the landscape, which may also contribute to the stability of the degraded system (Ravi et al., 2010). In the shrub encroachment case of the Chihuahuan Desert, for example, the increased bare soil area creates longer pathways for wind and water erosion. At the same time, the increased bare soil and sparser vegetation reduce the ability of fires, which would inhibit shrub establishment and favor grassland regeneration, to move across the landscape (Ravi et al., 2010).

Climate Change and Land Degradation

Degradation can be triggered by climate change and variability, with droughts and shifts in aridity being especially important. Increased aridity can cause reductions in vegetation growth and health, especially in drylands, where vegetation productivity is strongly limited by moisture. Drying of the soils and declines in vegetation coverage and growth can cause increased soil erosion, especially by the wind. During the 20th century, over the central plains of the United States, for example, dust storm frequency peaked during the Dust Bowl drought in the 1930s and another persistent drought in the 1950s (figure 7.4) (Chepil et al., 1963; Gillette & Hanson, 1989). In these cases, the degradation was likely forced by the combined influence of both climatic (i.e., a drought) and human (i.e., poor land-use practices) factors, with recovery (at least in terms of wind erosion and dust storms) occurring within several years of the ending of these droughts.

The paleoclimate record also includes many examples of drought-induced land degradation. One clear example, with little or no connection to human activities, is the megadroughts that plagued much of the central plains of North America during the Medieval Climate Anomaly. These events caused significant vegetation mortality, resulting in the mobilization of currently stable dune systems (e.g., the Nebraska Sand Hills) and increased wind erosion, all recorded as increased aeolian sediment deposition in lakes and ponds (Forman et al., 2001; Hanson et al., 2010; Miao et al., 2007). In many cases, these megadroughts, and the associated landscape degradation, persisted for decades.

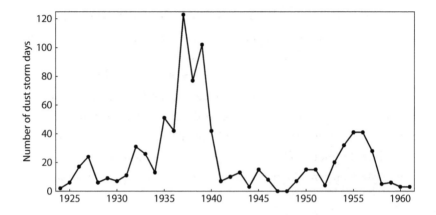

FIGURE 7.4 Observations of dust storm frequency (number of dust storm days per calendar year) for Dodge City, Kansas, from 1924 to 1961. Two major droughts (the 1930s Dust Bowl drought and the 1950s drought) occurred during this interval, and these events emerge clearly as enhanced periods of dust storm activity. *Source*: Data from Chepil et al., 1963.

Because of the strong connection between natural drought variability and land degradation, there are concerns that climate change will increase the susceptibility of ecosystems to land degradation and desertification in the coming decades. This is because many of the same regions that are already vulnerable to land degradation (e.g., drylands) will likely experience changes in hydroclimate (reduced precipitation and increased evaporative demand) that will further reduce water availability. Because of this, many dryland regions are expected to shift to more arid dryland categories by the end of the 21st century under high-warming (e.g., RCP 8.5) greenhouse gas–forcing scenarios (Huang et al., 2016). The most widespread dry shifts are expected to occur in West Africa, southern Africa, the Middle East, the Mediterranean, and China, regions that are already considered hot spots of degradation in several global assessments. Drylands are expected to shift to slightly wetter dryland categories in only a few locations, mostly in Australia and extratropical South America. Observed trends (Greve et al., 2014) over the latter half of the 20th century (1948–2005) are generally consistent with these projections, including increases in aridity that appear especially widespread over Africa, Asia, the Mediterranean, and the Middle East. This suggests that the climate changes expected to make these regions more susceptible to degradation may already be manifesting.

One factor that may ameliorate, or even counteract, climate change–induced drought stress and desertification is the increasing carbon dioxide (CO_2) concentrations in the atmosphere. Higher atmospheric CO_2 concentrations decrease drought stress on plants, allowing them to maintain higher levels of carbon assimilation and productivity even as moisture availability declines. This effect may be especially important for dryland ecosystems, where vegetation productivity is strongly limited by water. Indeed, greening trends in many global drylands have been observed in recent decades, and many of these trends have been attributed, at least in part, to increasing atmospheric CO_2 concentrations. One analysis of satellite vegetation records (Donohue et al., 2013) estimated that increases in atmospheric CO_2 from 1982 to 2010 are the primary driver of an 11 percent increase in green vegetation coverage in warm and arid regions of the world. A separate meta-analysis of CO_2 fertilization experiments (Lu et al., 2016) found that dryland sites under elevated CO_2 experienced about a 17 percent increase in soil water availability. Observations therefore support some potential ameliorating effects from increasing CO_2, although substantial uncertainties exist regarding the current and future importance of this effect, especially in light of expected changes in hydroclimate.

Land Degradation Feedbacks Within the Climate System

The important role of the land surface and land-cover feedbacks in drought dynamics has been discussed in previous chapters, and these mechanisms can act as positive feedbacks to further amplify land degradation. Initial losses of vegetation and ecosystem services at the surface can act in various ways to suppress regional precipitation and increase aridity. This increased aridity further suppresses plant and ecosystem productivity, creating a vicious cycle between the surface and atmosphere that acts to reinforce the degraded state. Even though there are still substantial uncertainties in the strengths of these feedbacks over different regions, the fundamental physics are well understood, and there are a variety of events where these processes are hypothesized to have contributed to land degradation. These mechanisms have been discussed previously in specific case studies (e.g., the Green Sahara, the Dust Bowl drought, and the Sahel drought), but here we briefly revisit them.

Increased bare soil exposure, a common consequence of land degradation, increases the surface albedo and reflection of shortwave energy, reducing net energy availability at the surface and top of the atmosphere. This induces anomalous subsidence and increases stability in the atmosphere, inhibiting precipitation and convection. Reductions in vegetation cover also reduce evapotranspiration from the surface; this occurs, for example, when leaf area declines (reducing the evaporative surface area) and when deeper-rooted vegetation (e.g., trees) is removed or replaced by vegetation with shallower roots, cutting off access to deeper soil moisture pools. With less evapotranspiration, less moisture is supplied to the atmosphere and boundary layer, leading to declines in precipitation recycling and moist static energy availability. Finally, vegetation losses affect surface roughness. Rough surfaces, such as the canopy of a forest, induce mechanical turbulence as wind flows past and can create pockets of convergence and rising motion, stimulating precipitation. Removal of vegetation reduces surface roughness, thereby reducing this effect.

The important role of mineral dust aerosols in the climate system is being increasingly recognized, especially from the perspective of land degradation feedbacks. Wind erosion commonly increases with vegetation losses from land degradation, especially in drylands. Globally, dust aerosols are produced by natural sources (mostly dry lake beds) and human land-use change, with anthropogenic sources accounting for ~25 percent of global emissions (Ginoux et al., 2012). In addition to representing a loss of nutrients from the soil, these mineral dust aerosols affect the energy balance at the surface and in the atmosphere and interact with clouds. Dust aerosols are strongly reflective in the shortwave part of the electromagnetic spectrum and also are effective absorbers of longwave energy. On balance, the shortwave effects usually dominate over the longwave effects, resulting in a net reduction in energy availability at the surface and in the atmosphere and thereby suppressing precipitation. Dust aerosols can further act as cloud condensation nuclei, inhibiting precipitation when they increase cloud lifetimes by dispersing nucleation across many more nuclei. The result is more, but smaller, cloud droplets that are less effective at precipitating.

Land surface, vegetation, and dust aerosol feedbacks have been cited as amplifiers of land degradation in a variety of studies and systems. These include the decline and collapse of the Green Sahara at the end of the mid-Holocene

(Tierney et al., 2017), the Sahel drought in the 1970s (Zeng et al., 1999), and the Dust Bowl drought of the 1930s (B. Cook et al., 2009; Schubert et al., 2004), all discussed in detail in previous chapters. Dust-climate feedbacks are also hypothesized to have influenced desertification in Mongolia and China (Xue, 1996) and in the central plains during the Medieval Climate Anomaly megadroughts (B. Cook et al., 2013). With climate change, these feedbacks are likely to play a continually important role, amplifying the expected greenhouse gas–forced drying in dryland and desert regions if vegetation coverage declines (Zeng & Yoon, 2009).

Mitigation of Land Degradation

The causes, manifestations, and impacts of land degradation are complex, making it difficult to develop robust projections of land degradation for the future. Climate change will likely play a central role in increasing the susceptibility of many regions to degradation, especially drylands regions. Beyond this, degradation risk will also depend on changes in regional human populations (e.g., shifts in population size or density and migrations that change demand for land), dietary changes (e.g., increased meat consumption), global demand for various commodities (e.g., palm oil and biofuels), and policies at different levels that discourage (or encourage) activities that will contribute to land degradation. Despite the seeming irreversibility of land degradation in many regions, however, various strategies have been developed to ameliorate or even reverse some aspects of land degradation.

One commonly deployed remediation method is the introduction of erosion control measures on the landscape (discussed in detail in D'Odorico et al., 2013). These can take a variety of forms and often require different techniques depending on whether wind or water erosion is the dominant concern. For example, artificial structures can be installed to improve routing of surface runoff to avoid the most extreme forms of water erosion (e.g., gully or sheet erosion) at the surface. Other techniques, such as installing more-permeable surfaces and replanting vegetation, can be deployed to improve infiltration and reduce total runoff. Lines of vegetation (windbreaks or shelterbelts) installed upwind of erodible fields can serve as momentum sinks to reduce wind erosion, and fields can be further protected from the wind by mixing crop residue into the soil after harvest. Overgrazing can be ameliorated by implementing grazing restrictions, including seasonal exclosures that prevent grazing of certain areas at certain times of year and allow an opportunity for the vegetation to recover. Often, however, heavily degraded lands will require active replanting of native vegetation to replace what was lost previously. For saline soils, active leeching through the addition of low-salinity water can help remove dissolved salts. Where salinization occurs because of waterlogged soils and high water tables (i.e., groundwater-induced salinization), drainage can be improved to allow this water to drain out of the soils. In many regions, however, salinization may be effectively permanent: for example, in areas where sufficient low-salinity water is not available or where very shallow, very saline water tables are difficult to drain.

Ultimately, land degradation has no singular cause and often involves many different factors (e.g., loss of vegetation and declines in soil fertility). No method of remediation will be equally suitable in all cases, and often multiple measures will be required to restore full ecosystem functionality. Further, in many cases, it is unlikely that recovery will happen spontaneously following removal of the primary causes of degradation (e.g., overgrazing), so direct and active human intervention will be required. It is therefore critical to have full stakeholder engagement and participation by local populations (D'Odorico et al., 2013), with both public and private institutions providing the necessary resources and incentives to improve resilience and remediation capacity.

EIGHT

Groundwater and Irrigation

Technological innovations over the course of the 20th century have increased human resilience in the face of many environmental challenges, including drought. Foremost among these have been the exploitation of groundwater and the expansion and intensification of irrigation. These innovations have increased the exploitable water resources available to farmers, improved agricultural yields, and enabled the expansion of croplands into regions previously too water limited for agriculture. Despite the clear advantages to agriculture and societies provided by groundwater exploitation and irrigation, however, there are significant concerns regarding the future sustainability of these practices in the face of climate change, increased demand, and overexploitation. Here, I provide an overview of groundwater exploitation and irrigation, the current state of these practices in the modern world, and the broader concerns regarding their sustainability in the future.

Groundwater

Groundwater is water stored in underground **aquifers**, formations of permeable rock and unconsolidated material, such as gravel and sand. Globally, it is the single largest store of freshwater, accounting for one-third of all human freshwater withdrawals (Taylor et al., 2013). Agriculture is the single largest use of groundwater, accounting for ~70 percent of total groundwater withdrawals (Margat & van der Gun, 2013), and groundwater supplies an estimated 36 percent, 42 percent, and 27 percent of the water used for domestic, agricultural, and industrial purposes, respectively (Taylor et al., 2013). Worldwide, ~2 billion people use groundwater as their primary water source, and groundwater is especially important in semiarid and lower-income countries with limited surface water resources (Taylor et al., 2013).

In many ecosystems, groundwater plays an important role as a source of sustained flow to rivers, lakes, and wetlands. This supply is especially important for streamflow during droughts or seasons with little rainfall (Taylor et al., 2013). Plants with deep enough roots can use groundwater, and, in wetlands and riparian zones in semiarid regions, many plants termed *phreatophytes* are dependent on this water source for survival during periods of insufficient precipitation (Naumburg et al., 2005). Much of the human-exploited groundwater comes from large aquifer systems that serve as the primary source of freshwater in many important agricultural regions (figure 8.1). Examples include the High Plains (Ogallala) Aquifer in the central United States, the Central Valley Aquifer in California, the Great Artesian Basin in Australia, the Indus Basin in India, and the North China Plain in eastern China. Between 600 and 1,100 km³ of groundwater are withdrawn every year (Siebert et al., 2010). The three countries that extract the most groundwater (based on 2010 estimates) are India (251 km³ yr⁻¹), China (111.95 km³ yr⁻¹), and the United States (111.70 km³ yr⁻¹) (Margat & van der Gun, 2013).

Natural recharge and replenishment of groundwater occurs from both direct diffuse infiltration of rainfall and more focused recharge from surface waters, such as streams and wetlands (Taylor et al., 2013). Focused recharge is particularly important in many semiarid regions where ephemeral lakes and other temporary water bodies form during occasional, but intense, rainfall events (Taylor et al., 2013). Broadly, groundwater recharge depends primarily on the local climate, which sets the overall moisture balance in a given region. In figure 8.1, for example, aquifers with the highest recharge rates (dark blue) are generally in areas with more mesic climates than are those with the lowest recharge rates (light blue). In the regions with less mesic climates, variability in climate can cause substantial variability in recharges rates from one year to the next. Land cover and vegetation growth affect recharge through the influence of the vegetation on evapotranspiration, canopy interception, and energy partitioning at the surface. These processes affect the amount of water available seasonally for runoff and, ultimately, groundwater recharge. Conversion of grassland to agriculture, for example, can increase groundwater recharge because losses from evapotranspiration are reduced during fallow periods between crop rotations when vegetation coverage is low (e.g., Scanlon et al., 2005). The soils and geology further influence the infiltration capacity of the soil (i.e., how much water reaches the aquifer) and the storage capacity of the aquifer itself.

Global diffuse recharge rates are estimated to be between 12,000 and 15,000 km^3 yr^{-1}, equivalent to about 30 percent of the available renewable freshwater resources on Earth (Siebert et al., 2010; Taylor et al., 2013). Studies using stable isotopes indicate that most aquifers have most recently experienced net recharge during cooler past climates, including glacial periods in the Pleistocene and the early part of the Holocene interglacial (5,000 years ago and earlier) (Taylor et al., 2013). Groundwater that was recharged during these past periods is typically referred to as *fossil groundwater*, highlighting the fact that these resources are largely nonrenewable on timescales of relevance for human societies because the current rates of withdrawal are significantly higher than modern recharge rates. There are an estimated ~22 million km^3 (16–30 million km^3, when uncertainties related to porosity estimates are accounted for) of fossil groundwater (here, defined as more than 100 years old) in the upper 2 km of continental crust (Gleeson et al., 2016). In especially dry regions, such as deserts, groundwater may be over 1 million years old, whereas in more humid places with higher recharge rates groundwater may be only months old (Gleeson et al., 2016). Very little groundwater (~350,000 km^3) is classified as modern (less than 50 years old), though even this relatively small pool is larger than most other terrestrial sources of freshwater, including lakes, rivers, and the upper soil (Gleeson et al., 2016). Modern groundwater is generally more susceptible to climate variations and trends in the hydrologic cycle through the impact of climate variability on recharge rates (Gleeson et al., 2016). For most of the globally important aquifers, however, modern recharge rates are small relative to storage, meaning that groundwater is withdrawal limited and effectively nonrenewable.

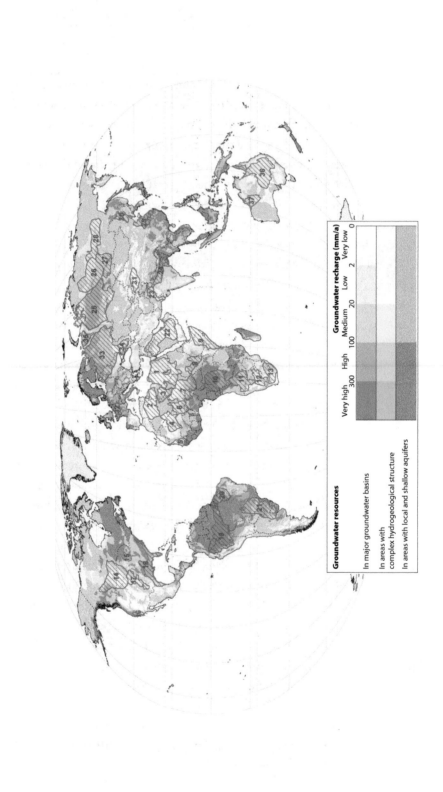

Groundwater resources

In major groundwater basins

In areas with
complex hydrogeological structure

In areas with local and shallow aquifers

Groundwater recharge (mm/a)

Very high	High	Medium	Low	Very low
300	100	20	2	0

FIGURE 8.1 The global distribution of major groundwater aquifer systems, including estimated natural recharge rates. In many agriculturally important regions, groundwater from these aquifers supplies a large portion of the water used to irrigate crops. These include the Central Valley Aquifer in California, the High Plains (or Ogallala) Aquifer in the central United States, and the Indus Basin in India. Groundwater recharge rates depend primarily on local climate, with higher rates in more mesic regions. In many aquifers, however, groundwater is effectively nonrenewable (often referred to as *fossil groundwater*) because extraction rates greatly exceed natural recharge rates, leading to significant depletion of these resources.

1. Nubian Aquifer System (NAS),
2. Northwest Sahara Aquifer System (NWSAS)
3. Murzuk-Djado Basin
4. Taoudeni-Tanezrouft Basin
5. Senegalo-Mauritanian Basin
6. Iullemeden-Irhazer Aquifer System
7. Chad Basin
8. Sudd Basin (Umm Ruwaba Aquifer)
9. Ogaden-Juba Basin
10. Congo Intracratonic Basin
11. Northern Kalahari Basin
12. Southeast Kalahari Basin
13. Karoo Basin

14. Northern Great Plains/Interior Plains Aquifer
15. Cambro-Ordovician Aquifer System
16. California Central Valley Aquifer System
17. High Plains-Ogallala Aquifer
18. Gulf Coastal Plains Aquifer System
19. Amazonas Basin
20. Maranhao Basin
21. Guarani Aquifer System
22. Arabian Aquifer System
23. Indus Basin
24. Ganges-Brahmaputra Basin
25. West Siberian Artesian Basin
26. Tunguss Basin

27. Angara-Lena Artesian Basin
28. Yakut Basin
29. North China Plain Aquifer System
30. Songliao Basin
31. Tarim Basin
32. Parisian Basin
33. East European Aquifer System
34. North Caucasus Basin
35. Pechora Basin
36. Great Artesian Basin
37. Canning Basin

Source: Large Aquifer Systems of the World, prepared by the WHYMAP project in cooperation with Jean Margat, https://www.whymap.org.

Irrigation

Irrigation is the supply of water used to assist in the growth and production of crops. In the modern era, irrigation accounts for ~60 to 70 percent of global freshwater withdrawals and up to 80 percent of total freshwater consumption (Siebert et al., 2015; Wada et al., 2013). Irrigation is the single largest human water use, drawn from both surface reservoirs and groundwater, and its importance for late 20th-century agriculture is difficult to overstate. Irrigation can double crop yields compared to the situation where the same crops are grown on rain-fed systems (Siebert et al., 2015), and, in drier regions, irrigation is required because of insufficient precipitation. Only ~20 percent of global cultivated lands are irrigated, but crops grown on these lands account for ~40 percent of global food production (Siebert et al., 2005).

Information on historical and modern irrigation rates is not readily available from most regions, so these datasets typically combine empirical observations of areas equipped for irrigation (AEI) and estimates of irrigation water demand/withdrawals (IWD) calculated from process-based hydrologic models. Irrigated areas expanded rapidly over the course of the 20th century, accelerating after ~1950 (figure 8.2). Global AEI was ~63 million ha (Mha) in 1900, expanded to ~111 MhA in 1950, and nearly tripled over the latter half of the 20th century to 306 Mha in 2005 (Siebert et al., 2015). Irrigation expansion was most rapid in the middle of the century (from the 1950s to 1980s), growing at about 5 percent per year before slowing to an expansion rate of less than 1 percent in more recent decades (Wada et al., 2013). The clear majority of AEI expansion has occurred in South and East Asia, regions that also account for most of global IWD (Siebert et al., 2015).

Global estimates for late 20th-century average IWD (including both surface and groundwater sources) range from ~2,000 to ~3,000 km^3 yr^{-1}, reflecting broad uncertainties in the underlying hydrologic modeling (Siebert et al., 2015; Wada et al., 2014). Withdrawal and use of irrigation water is strongly seasonal, typically peaking during local summer and the main growing season (figure 8.3). In many tropical and subtropical regions, where the growing season is less temperature limited (e.g., India and Southeast Asia), irrigation can also be significant during the winter dry season. Total annual irrigation

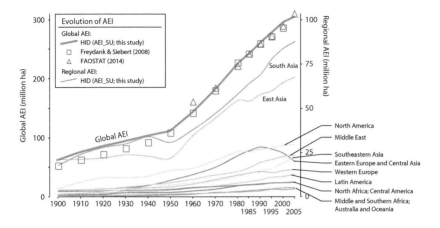

FIGURE 8.2 Global and regional changes in areas equipped for irrigation (AEI) over the 20th century from various data sources. Irrigated areas have expanded steadily in most regions, accelerating after 1950 in tandem with other agricultural improvements associated with the Green Revolution. Peak expansion rates occurred in the mid-20th century, before slowing in more recent decades. Globally, the overwhelming majority of irrigation is concentrated in South Asia and East Asia. *Source:* Reprinted from Figure 3, Siebert et al., 2015; figure also includes data from Freydank & Siebert, 2008, and Food and Agriculture Organization of the United Nations, 2016.

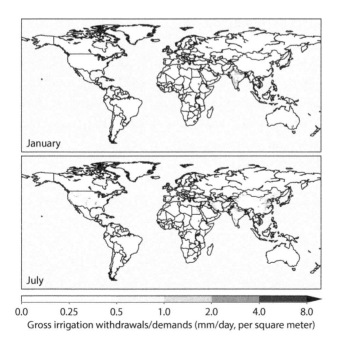

FIGURE 8.3 Estimates of gross irrigation withdrawals/demand for January and July 2005. In most regions, irrigation rates peak during the summer growing season, though many subtropical regions (e.g., India) also have substantial irrigation during the relatively warm dry season (boreal winter) when water demand peaks. *Source:* Data from Wada et al., 2014.

water consumption by crops is estimated at 1,277 km^3, with 545 km^3 (43 percent) originating as groundwater (Siebert et al., 2010). The contribution of groundwater to irrigation varies significantly across regions and countries and depends strongly on the local availability of surface versus groundwater resources. Approximately 38 percent of global irrigated lands are equipped to use groundwater (Siebert et al., 2010).

The difference between the much higher IWD (2,000–3,000 km^3 yr^{-1}) and the lower crop consumption rate (1,277 km^3 yr^{-1}) reflects the fact that not all irrigation water transported and applied to a field is used by plants. The fraction of water withdrawn and transported for irrigation that is used by plants is referred to as the scheme irrigation efficiency (Brower et al., 1989). This efficiency can be further subdivided into the conveyance efficiency (the efficiency of water transport to the field from the irrigation water source) and field application efficiency (the efficiency related to the application of water in the field). These efficiencies depend on a variety of factors, including local climate (e.g., evaporative demand in the atmosphere), crop type, permeability of soils, field slope, conveyance method, and irrigation method. Typical conveyance losses occur through evaporation and spills from open canals and through seepage or leaks from canals or pipes; losses are generally higher from canals than from closed pipes (Howell, 2003). Even in the most efficient applications, some proportion of irrigation is typically lost from the field via runoff (at the surface or belowground) or percolation of water below the root zone.

Irrigation methods fall into three broad categories. The most widespread methodology is surface irrigation, in which water is spread across fields and infiltrates into the soil through gravity. Rice paddies in Asia and almond groves in California are typical examples of agricultural systems that use surface irrigation. This method is most effective in areas with mild slopes, medium to low infiltration rates in the soils, and a large supply of available water, as field application efficiency is typically only ~60 percent (Brower et al., 1989). In sprinkler irrigation, water is distributed by high-pressure sprinklers, typically from a central pivot or moving platform with a typical field application efficiency of ~75 percent (Brower et al., 1989). This type of irrigation is common over wheat fields in the central United States (e.g., Kansas). The third form, drip irrigation, delivers water directly to or near the roots of plants, minimizing losses from surface evaporation and runoff. This is the most efficient method, with a field application efficiency up to about 90 percent.

Groundwater and Irrigation Interactions with the Earth System

Groundwater, irrigation, and the climate system interact through a variety of pathways, modulated by both natural environmental processes and human activities. To satisfy irrigation and other human water demands, water is withdrawn from surface reservoirs (e.g., rivers and lakes) and groundwater reservoirs. This water is then redistributed to plants and soils and often flows back to different surface reservoirs and aquifers, all movements sensitive to climate, land cover, and agricultural practices. Such perturbations to the hydrologic cycle are not passive, however, and can, in turn, influence or act as feedbacks to other physical and biological processes in the Earth system. For example, evapotranspiration from irrigated crops can send large quantities of water from the surface back to the atmosphere, influencing the surface energy balance and local weather and climate. High rates of withdrawal can cause groundwater depletion, contributing to land subsidence and loss of storage capacity, saltwater intrusion along the coast, and sea-level rise. Water cycle responses to irrigation and groundwater extraction are thus dynamic, and any comprehensive analysis of these processes must consider the array of interactions between them and the Earth system.

Groundwater Depletion

Groundwater resources are overwhelmingly dominated by older fossil groundwater that is effectively nonrenewable on timescales most relevant for human activities (i.e., years to decades). Such aquifers are storage rather than recharge-flux dominated because withdrawal rates greatly exceed natural recharge (Taylor et al., 2013). Depletion of groundwater is an issue of considerable international concern, especially in semiarid and arid areas where surface water resources are limited and in regions with exceptionally high withdrawal rates (e.g., because of intense irrigation requirements)

(Taylor et al., 2013). One major impediment to accurately quantifying groundwater availability, however, is the paucity of accurate, direct measurements of water storage within many aquifers (Richey, Thomas, Lo, Famiglietti et al., 2015). As a result, most large-scale assessments of groundwater depletion instead track changes in, rather than absolute storage of, groundwater in aquifers. This information can be determined using mass balance approaches and also using data from the Gravity Recovery and Climate Experiment (GRACE) satellites. These satellites measure changes in local gravitational anomalies related to mass changes at the surface. Over many terrestrial regions, these mass changes coincide primarily with changes in terrestrial water storage, which includes water stored in vegetation, in soils, at the surface in lakes and rivers, and in aquifers. GRACE is not capable of estimating the actual amount of water stored at a given location, but it can provide relatively accurate estimates of whether storage is declining or increasing.

GRACE has documented significant groundwater depletion in many major aquifer systems over its period of operation (2002–2017) (Famiglietti, 2014). Some of the highest rates of depletion are in agriculturally intensive regions of Asia, areas with some of the highest irrigation rates in the world. These include northwestern India, with rates of -17.7 km^3 yr^{-1} from 2002 to 2008, and the North China Plain, with rates of -8.3 km^3 yr^{-1} from 2003 to 2010. A recent global assessment (Richey, Thomas, Lo, Reager et al., 2015) concluded that 13 of the 37 largest aquifers in the world can be classified as stressed, meaning that water is being withdrawn at a rapid pace but the aquifer is receiving little to no recharge (figure 8.4). They identified the three most stressed aquifers as the Arabian Aquifer, the Indus Basin, and the Murzuk-Djado Basin in North Africa, regions that are relatively dry and/or have intense agriculture and irrigation. In some regions, however, these large-scale assessments can mask important subaquifer variability. Most of the depletion of the High Plains Aquifer, for example, is occurring in the southern and central part of the plains, whereas the northern part has experienced very little depletion (Scanlon et al., 2012). Additionally, GRACE launched only in 2002, and estimates of groundwater depletion during prior decades can be difficult to quantify because of the paucity of available data. For many areas with longer historical records, however, it is clear that groundwater depletion has been ongoing for decades. Over the

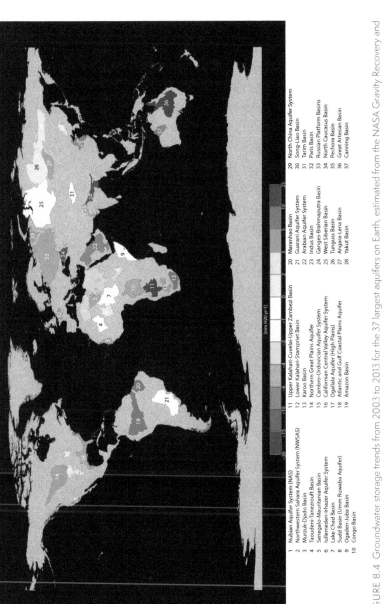

FIGURE 8.4 Groundwater storage trends from 2003 to 2013 for the 37 largest aquifers on Earth, estimated from the NASA Gravity Recovery and Climate Experiment (GRACE) satellites. Aquifers with negative trends (yellow to red in color) are considered stressed because the current rates of withdrawal far exceed the natural rates of recharge. If this stressed condition continues, the accessible water in these aquifers will eventually be fully depleted. Lack of information regarding the actual amount of water in many aquifers, however, makes it difficult to predict when this would occur. Of these 37 aquifers, 13 are considered to be significantly distressed with little to no recharge at all, a situation likely to undermine regional water security.

Source: UC Irvine/NASA/JPL-Caltech, https://www.jpl.nasa.gov/news/news.php?feature=4626.

1 Nubian Aquifer System (NAS)
2 Northwestern Sahara Aquifer System (NWSAS)
3 Murzuk-Djado Basin
4 Taoudeni-Tanezrouft Basin
5 Senegalo-Mauritanian Basin
6 Iullemeden-Irhazer Aquifer System
7 Lake Chad Basin
8 Sudd Basin (Umm Ruwaba Aquifer)
9 Ogaden-Juba Basin
10 Congo Basin

11 Upper Kalahari-Cuvelai-Upper Zambezi Basin
12 Lower Kalahari-Stampriet Basin
13 Karoo Basin
14 Northern Great Plains Aquifer
15 Cambro-Ordovician Aquifer System
16 Californian Central Valley Aquifer System
17 Ogallala Aquifer (High Plains)
18 Atlantic and Gulf Coastal Plains Aquifer
19 Amazon Basin

20 Maranhao Basin
21 Guarani Aquifer System
22 Arabian Aquifer System
23 Indus Basin
24 Ganges-Brahmaputra Basin
25 West Siberian Basin
26 Tunguss Basin
27 Angara-Lena Basin
28 Yakut Basin

29 North China Aquifer System
30 Song-Liao Basin
31 Tarim Basin
32 Paris Basin
33 Russian Platform Basins
34 North Caucasus Basin
35 Pechora Basin
36 Great Artesian Basin
37 Canning Basin

United States, for example, irrigation for agriculture has driven most of the depletion from the Central Valley Aquifer in California and the High Plains Aquifer in the central United States. From 1900 to 2008, depletion from these aquifers accounted for ~50 percent of total groundwater depletion in the United States, equivalent to a loss of ~400 km³ over the latter half of the 20th century (Scanlon et al., 2012).

Groundwater depletion in many regions typically intensifies during major drought events when surface water resources begin to decline. Such a situation occurred in the early 21st century over California, where a series of successive drought events forced the state to rely increasingly on groundwater to meet demands (Xiao et al., 2017). Using mass balance approaches and GRACE, Xiao et al. (2017) assessed changes in groundwater storage over the Central Valley. They demonstrated that during two recent drought periods in California (2007–2009 and 2012–2016), groundwater declined over the Central Valley (one of the main agricultural regions in the state) by ~5.5 km³ yr⁻¹ and 10–11 km³ yr⁻¹, respectively. The higher losses during the more recent drought are due to a variety of factors, including greater crop water use and reduced inflows, highlighting the importance of both physical and social factors. Mass balance–based estimates place total net losses of groundwater from April 2002 to the end of the most recent drought in September 2016 at ~20 km³, with much more severe losses (nearly triple) estimated by the GRACE satellites. These represent substantial declines in storage, being of the same order of magnitude as the maximum capacity of Lake Mead (~36.6 km³ in 2010), the largest and single most important surface reservoir in the western United States (National Park Service, 2015).

Land-use factors and irrigation can significantly affect local groundwater recharge rates, in some cases ameliorating depletion or even causing net recharge. Dry season irrigation in South Asia increases recharge and total groundwater storage during the subsequent monsoon season, and similar cases have been documented in China, the Middle East, Pakistan, and the Central Valley of California (Taylor et al., 2013). Some of the reduced depletion in the northern part of the High Plains Aquifer has been attributed to infiltration from irrigation drawn from surface water (Scanlon et al., 2012). Shifts in land use from natural savanna to managed crops also increased groundwater in the Sahel during the late 20th century, a consequence of increased runoff in ponds that created areas of focused recharge (Leblanc et al., 2008). Similar shifts from native vegetation to croplands also increased groundwater recharge in Australia and the U.S. Southwest earlier in the 20th century (Taylor et al., 2013). One approach to address groundwater depletion locally is groundwater banking, whereby water from the surface is purposefully stored in aquifers. This can be accomplished through either direct injection or surface spreading (where water is pooled at the surface and allowed to infiltrate) (Maliva, 2014). The fundamental idea is to store excess water during seasons or years of surplus belowground for withdrawal during later periods when needed. Groundwater banking has been implemented to a limited extent in the semiarid western United States, including California, Nevada, and Arizona (Maliva, 2014) and has some advantages over storage in surface reservoirs, such as reduced losses from evapotranspiration (Taylor et al., 2013). However, successful groundwater banking and use of the banked water requires favorable climatic and geologic conditions, as well as clear management plans governing inputs and withdrawals (Maliva, 2014).

Beyond the loss of water resources, groundwater depletion can have a variety of other impacts on the Earth system, locally and globally. In regions of exceptionally severe depletion, loss of water from an aquifer can cause subsidence and settling of the land surface. Subsidence damages surface infrastructure (e.g., roads and bridges) and causes a loss of porosity within the aquifer that permanently limits groundwater recharge in the future. Parts of the San Joaquin Valley in California, for example, subsided as much as 85 cm from 2007 to 2010 due to groundwater pumping, translating to a loss of groundwater storage equivalent to 5–9 percent of the total groundwater used in that period (Smith et al., 2017). Depletion over the last several decades may have even caused some areas of California to sink by as much as 30 feet (Null, 2017). Subsidence related to groundwater pumping has also been observed in the North China Plain (Hwang et al., 2016), the Mekong Delta (Erban et al., 2014), and many other locations around the world (Galloway & Burbey, 2011). An additional challenge in coastal aquifers is seawater intrusion, which degrades water quality. Risk factors for seawater intrusion are complicated and include coastal topography, groundwater recharge rates, local sea-level changes, and withdrawal rates (Taylor et al., 2013). Studies indicate, however, that coastal aquifers are likely much more vulnerable to seawater intrusion from groundwater extraction than from sea-level rise across a wide diversity of contexts (G. Ferguson & Gleeson, 2012). Groundwater extraction–induced seawater intrusion has been documented in many regions, including Asia, the eastern Mediterranean, and the United States (Barlow & Reichard, 2010; Taniguchi, 2011; Yakirevich et al., 1998).

Groundwater extraction is also a minor, albeit significant, contributor to global sea-level rise. Because of the low rates of turnover of groundwater in many aquifers, this extracted water is effectively removed from annual- to decadal-scale cycling in the Earth system. When this water returns to the modern water cycle through, for example, irrigation of crops, it effectively represents a new, external input to the climate system and modern water cycle. Most of modern sea-level rise itself is caused by thermal expansion and the melting of land ice because of climate change. Estimates of the groundwater contribution vary, primarily due to uncertainties in different estimates of global groundwater depletion. Konikow (2011), for example, estimated that from 1900 to 2008 groundwater depletion contributed a total of 12.6 mm of sea-level rise, or about 6 percent of the total sea-level rise that occurred over that period. For a more recent decade (2000–2008), they estimate that the groundwater contribution has increased to a rate of 0.40 mm yr^{-1}, equivalent to ~13 percent of the total observed rate of 3.1 mm yr^{-1} of sea-level rise in that interval. Other estimates suggest higher overall rates of groundwater depletion—and thus larger contributions to sea-level rise. Wada et al. (2012), for example, calculated a 0.57 mm yr^{-1} contribution of groundwater depletion to sea-level rise for the year 2000 and projected this will likely increase to 0.82 mm yr^{-1} by 2050.

Irrigation and Groundwater Interactions with Climate

Much of our understanding of how irrigation and groundwater interact with the atmosphere and climate system is based on work using coupled climate models. Although climate models have historically had very shallow soils (typically extending to depths of several meters, at most), recent improvements to the models have highlighted the potentially important role of deeper moisture pools associated with groundwater. When included in models, such groundwater interactions can significantly influence land-atmosphere fluxes, affecting the growth of the boundary layer, soil and air temperatures, and even precipitation (I. Ferguson & Maxwell, 2010; Maxwell & Kollet, 2008). Water table depths of between 2 and 5 m appear to exert an especially strong influence over energy fluxes at the land surface in models (Maxwell & Kollet, 2008). Even now, however, most climate models do not include explicit treatments of groundwater processes, and there are thus still large uncertainties regarding how groundwater affects climate across regions and in the context of climate change.

The more direct and well understood pathway for groundwater to influence climate and weather is through irrigation. Along with deforestation, the expansion of agriculture, urbanization, and other land-use practices, irrigation represents a major anthropogenic modification of the land surface. Irrigation impacts weather and climate primarily by shifting the surface energy balance to favor latent over sensible heating in regions where evapotranspiration is limited by the availability of moisture at the surface rather than by evaporative demand (Puma & Cook, 2010). When included in climate models, irrigation typically cools soil and air temperatures, increases surface air humidity, and increases cloud cover and precipitation (e.g., Boucher et al., 2004; B. Cook, Shukla et al., 2015; Lobell et al., 2009; Puma & Cook, 2010; Sacks et al., 2009). Typically, the most robust model responses are highly localized near the regions of most intense irrigation. An example for the late 20th century from the Goddard Institute for Space Studies (GISS) climate model is shown in figure 8.5. This figure shows differences in select variables between a simulation with irrigation included (using the dataset in figure 8.3) and an identical control simulation without irrigation. Irrigation causes widespread cooling (2 m air temperatures) across the most intensely irrigated regions of North America, Asia, and the Middle East, with concomitant increases in specific humidity, cloud cover, and precipitation in many of the same regions. Similar irrigation-induced climate shifts have been documented in other models as well (Boucher et al., 2004; Kueppers et al., 2007; Lo & Famiglietti, 2013; Lobell et al., 2009; Sacks et al., 2009), suggesting that irrigation impacts are relatively robust and constitute a significant component of anthropogenic land-use forcing of climate (B. Cook, Shukla et al., 2015).

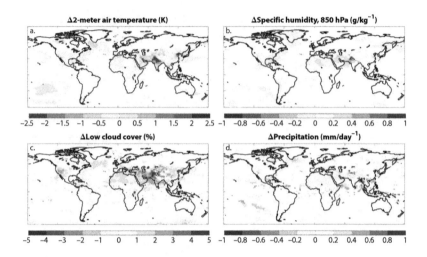

FIGURE 8.5 Model climate responses to late 20th century irrigation in the Goddard Institute for Space Studies (GISS) climate model: air temperature at 2-meter height above the surface, specific humidity at 850 hectopascals (hPa), low cloud cover, and precipitation. Irrigation causes cooling of near-surface air temperatures and increased humidity, cloud cover, and precipitation, with the largest responses collocated over regions with the most intense irrigation rates. These include North America, the Middle East, and Asia. Broadly, model responses are similar to what has been observed, especially for temperature. However, significant uncertainties remain in the magnitude and significance of the responses of many variables, especially precipitation, to irrigation. *Source:* Figure 6, B. Cook, Shukla et al., 2015.

Evidence for significant irrigation impacts on climate and weather in observations, however, is more mixed compared to the responses in models, with the clearest effect emerging in surface air and soil temperatures. The cooling effect of irrigation has been observed over the Central Valley of California (Bonfils et al., 2007; Lobell & Bonfils, 2008; Lobell et al., 2008), the central plains of the United States (Lobell et al., 2008; Mahmood et al., 2006), India (Roy et al., 2007), and China (Han & Yang, 2013; Shi et al., 2014; Xu et al., 2017), where it is seen most clearly in daytime temperatures and heat extremes, phenomena where cooling from increased evapotranspiration and cloud cover are likely to be most important (e.g., Lobell & Bonfils, 2008; Lobell et al., 2008). Observational evidence for significant irrigation impacts on precipitation are more mixed. Several studies have found evidence that increased precipitation over the southern plains (Barnston & Schickedanz, 1984), central plains (DeAngelis et al., 2010), and Midwest (Alter, Fan et al., 2015) of the United States is associated with high irrigation rates. There is also some observational evidence for irrigation enhancement of precipitation in East Africa (Alter, Im, & Eltahir, 2015). The efficacy of this effect, however, may depend on favorable atmospheric conditions that promote convergence and uplift (e.g., Barnston & Schickedanz, 1984). These greater uncertainties regarding the precipitation responses to irrigation are unsurprising given the still large uncertainties surrounding soil moisture–precipitation feedbacks (Seneviratne et al., 2010). Even more uncertain are the potential impacts of irrigation on larger-scale dynamics and regional teleconnections, such as the Asian monsoon, questions that have been explored almost exclusively with models. Vrese et al. (2016), for example, found that irrigation over monsoon Asia acted to enhance precipitation over East Africa. And there are several studies (Guimberteau et al., 2012; Shukla et al., 2014) that suggest the intensification of irrigation over India during the 20th century may be responsible, at least in part, for the recently observed weakening of the monsoon. Although intriguing, without either more realistic treatments of irrigation in models or better observational evidence, these results remain highly speculative.

Climate Change and the Future of Irrigation and Groundwater Resources

Climate change will alter the hydrologic cycle and hydroclimate in various ways (see chapter 5), many of which are likely to affect the sustainability of groundwater extraction and irrigation in the future. The impact of climate change on these resources and practices will be complicated, however, by the additional influence of human activities. For example, groundwater levels and recharge rates are likely to respond directly to changes in climate in some regions because of changes in precipitation amount, type, and intensity. In other areas, groundwater may decline as climate change increases drought risk or the irrigation requirements for agriculture. In such cases, the impact of climate change on groundwater will be indirect, modulated by human responses to climate change and policies that govern agriculture and water resources.

Groundwater in relatively shallow, smaller aquifers with higher modern recharge rates will be most sensitive to climate change, affected by shifts in precipitation (including its amount, timing, and intensity), evapotranspiration, and snow dynamics (e.g. proportion of precipitation falling as rain versus snow, snowmelt timing, etc.) (Kløve et al., 2014). The impact of specific changes in hydroclimate will, however, be context dependent, modulated by the local climate, land cover, and other processes. For example, climate change will likely lead to a strong increase in precipitation intensity and a higher frequency of heavy rainfall events (O'Gorman, 2015). In semiarid regions, where groundwater recharge is most closely tied to focused recharge events from intense precipitation, such shifts may increase groundwater recharge (Bates et al., 2008). In more humid regions, however, increased rainfall intensity may decrease recharge rates because more water will be lost from increased surface runoff (Bates et al., 2008). Indeed, whether one accounts for changes in intensity, or not, can affect whether projections indicate declining or increasing groundwater recharge rates, even for the same change in total precipitation (Taylor et al., 2013).

In colder regions, changes in the timing and magnitude of groundwater recharge with climate change are likely to be highly sensitive to changes in snow and soil freezing. Snow cover and the fraction of precipitation falling as snow decline with warming, and the timing of the snowmelt season lengthens and shifts earlier in the year. This can lead to extended periods of runoff because the snow will no longer melt in a single pulse in the spring, as typically occurs for many regions (e.g., the Sierra Nevada in California) in the modern climate. Such a shift would increase groundwater recharge during the winter and earlier in the year (Mäkinen et al., 2008; Okkonen & Kløve, 2010; Veijalainen et al., 2010) but likely reduce recharge during the subsequent spring and summer months (Okkonen & Kløve, 2010). Reductions in soil frost with warmer soils can also increase infiltration, increasing groundwater recharge during winter in regions with significant liquid precipitation (Jyrkama & Sykes, 2007; Okkonen & Kløve, 2011).

For larger and deeper aquifers, especially those containing primarily fossil groundwater, climate change impacts are most likely to manifest as societies increase or decrease their rates of groundwater extraction in response to climate change. Such extractions will likely be dominated by shifts in irrigation demands and requirements. Globally, irrigated areas are not expected to expand much in the near future because of limited land and water availability (Faurès et al., 2002; Turral et al., 2011). The impacts of climate change

on IWD will thus be most relevant for locations already reliant on irrigation, increasing in regions where warming leads to precipitation declines or
increases in evaporative demand. Beyond this simple expectation, however,
are significant uncertainties in projections of IWD, including (1) the cal-
culation of IWD across various hydrologic models, (2) the magnitude and
direction of climate responses to increased greenhouse gas forcing across climate models, and (3) the ultimate emissions trajectory for the future. One
recent assessment (Wada et al., 2013) attempted to account for these various
uncertainties in a comprehensive way, calculating changes in IWD across
five hydrologic models, five climate models, and four greenhouse gas concentration scenarios. In RCP 8.5 (the typical high-emissions, high-warming
scenario), IWD increases on average by more than 25 percent by the 2080s in
the most intensely irrigated regions of Asia, Africa, Europe, and the United
States. In most cases, the increase in IWD is driven by increased evaporative
demand in a warmer atmosphere, even outweighing the effect of increased
precipitation in some regions. Unsurprisingly, IWD increases are reduced
for the scenarios with lower emissions and warming. One hydrologic model
(LPJml) did, however, project declining IWD in the future, in spite of the
projected climate changes. This model incorporates processes related to
the effect of carbon dioxide (CO_2) on crop and plant physiology, including the tendency for increased CO_2 concentrations to increase water-use
efficiency. In simulations using LPJml, this CO_2 effect dominates, leading to
net plant water savings and declines in IWD in spite of the climate changes.
Although encouraging, the magnitude and global importance of this effect
for surface hydrology are still highly uncertain.

Other factors unrelated to climate change will also affect the sustainability of groundwater resources and irrigation practices. These include changes
in populations and per capita water consumption, shifts in agricultural and
land-use practices, and the various laws and policies for managing existing water resources. Future sustainability will likely depend on improved
knowledge of how much water is available and economically extractable
in the most important aquifers, decisions about the best use of this water
(including during droughts when surface resources decline), and a better
understanding of how processes that might increase recharge (e.g., more
intense precipitation) can be exploited to increase groundwater resources

(Taylor et al., 2013). As in the past, water will remain a critical and challenging resource to manage in most regions of the world. But as demands increase with climate change, population growth, and agricultural intensification, and as previously reliable resources become insufficient for even current needs, improving our understanding how the complex interplay of physical and social processes will affect water availability in the coming decades will be more critical than ever.

Glossary

AQUIFER. A layer of permeable rock or unconsolidated materials (e.g., gravel and sand) that can store groundwater.

ARIDITY. A permanent or semipermanent state of dryness and low surface water availability, typically associated with regions where precipitation is low. Aridity is differentiated from *drought* because the latter represents transient (temporary) water deficits relative to the average baseline conditions within a specific region. On long timescales, however, the difference between the two can become less clear. For example, the Sahara Desert can transition between hyperarid and humid (Green Sahara) hydroclimate states that can persist for thousands of years. Similarly, climate change projections of increased drought risk in regions like the Mediterranean can also be viewed as a shift toward a more arid mean hydroclimate regime. A common measure of aridity is the *aridity index*, defined as the ratio of precipitation to potential evapotranspiration.

ATLANTIC MULTIDECADAL OSCILLATION (AMO). A multidecadal mode of climate variability associated with sea surface temperature variations in the North Atlantic. During warm phases, the AMO causes dry conditions across North America (especially in the central plains and Mississippi River valley) and northern tropical South America and wet conditions across West Africa.

BASELINE WATER STRESS. For a given location, the ratio of annual water withdrawals (demand) to the annual renewable supply (typically estimated as total annual runoff or streamflow).

BLUE WATER. Water contained in reservoirs, including rivers, lakes, wetlands, groundwater, and artificial impoundments (lakes created by dams). Runoff is the primary source of blue water, which represents the traditional sphere of human water resource management (contrast with *green water*).

CARBON DIOXIDE (CO_2) PHYSIOLOGICAL EFFECTS. Plants respond directly to increased atmospheric carbon dioxide (CO_2) concentrations by increasing their water-use efficiency, reducing water losses per unit of carbon assimilated. All else being equal, this would be expected to increase vegetation resilience to drought and diminish the drying effect of increased atmospheric temperature and aridity. However, higher CO_2 is also expected to increase photosynthesis, carbon assimilation, and overall biomass or carbon storage. This can lead to larger leaf areas and higher total plant water use, even as water-use efficiency increases. Ultimately, the responses of plants to atmospheric CO_2 and climate change (especially changes in temperature and precipitation) represent some of the largest uncertainties in climate model projections of drought.

CARBON STARVATION. A hypothesized vegetation mortality mechanism during drought, occurring when plant productivity and available carbohydrate reserves can no longer meet the requirements for basic metabolism. This is hypothesized to occur when stomatal closure in the face of moisture deficits limits gas exchange between the leaves and the atmosphere, reducing the availability of carbon dioxide for photosynthesis.

CHARNEY HYPOTHESIS. An important early hypothesis offered to explain the mechanism by which land degradation over the Sahel region would have caused or contributed to the Sahel drought that began in the 1960s. Charney focused on the effect of vegetation losses and land degradation on surface albedo and on the drying that would arise because of reductions in energy availability. Although subsequent work has shown that his mechanism/model was oversimplified, his hypothesis forms the basis for many subsequent studies investigating land-atmosphere interactions and drought over the region, including not only the mid-20th-century Sahel drought but also the collapse of the Green Sahara.

CLAUSIUS-CLAPEYRON RELATIONSHIP. The relationship between temperature and saturation vapor pressure. As temperature increases, the saturation vapor pressure increases exponentially (~7 percent per degree of warming). Over most land areas, where the supply of moisture at the surface is limited and may not be able to keep up with increased demand, this means that warming (e.g., from climate change) can cause reductions in relative humidity and increases in vapor pressure deficit, which can have ramifications for future hydroclimate and drought.

CLIMATE PROXIES. Naturally occurring archives that can be used to reconstruct or infer information on past climate variability (e.g., temperature, precipitation, atmospheric composition, etc.). Examples include tree rings, ice cores, corals, and sediment cores from lakes and oceans.

CONSUMPTION AND WITHDRAWAL. *Withdrawal* refers to the total amount of water taken from a source for any purpose; some withdrawals may return to the original source for use again (e.g., as runoff from an irrigated agricultural field). *Consumption* refers to water that is used and no longer available (~50 to 60 percent of global water withdrawals).

DESERT. A region with little precipitation or extremely low values of the aridity index. Vegetation in deserts tends to be sparse, with low levels of productivity throughout the year. Major deserts typically occur in four types of regions: (1) in the subtropics, beneath the descending branches of the Hadley circulation; (2) on the leeward side of mountain ranges (rain shadows); (3) in the center of continents far from available moisture sources; and (4) on the west coasts of continents near areas of intense oceanic upwelling.

DROUGHT. An anomalous moisture deficit relative to some (often arbitrarily defined) "normal" baseline. Droughts are typically classified based on where in the hydrologic cycle they manifest. Most droughts are caused by anomalously low precipitation (*a meteorological drought*). If these deficits persist for a long enough time, they can propagate into the soil moisture (*an agricultural drought*) and into runoff, streamflow, surface reservoirs, and groundwater (*a hydrological drought*). For these latter two categories of drought, other processes beyond precipitation (e.g., evapotranspiration, vegetation, and soil properties) can play important roles. Importantly, impacts on ecosystems, agriculture, and societies begin to emerge most strongly during agricultural and hydrological droughts.

DUST BOWL. A major, multiyear drought event during the 1930s that affected large areas of the United States, especially over the central plains. This drought was notable for its spatial extent and impacts on people and the landscape through widespread crop failures, land degradation, and dust storm activity. The drought was likely initiated by cold sea surface temperatures in the eastern tropical Pacific (a La Niña event) and warm sea surface temperatures in the tropical Atlantic, amplified by feedbacks from soil moisture deficits, vegetation losses, and dust aerosols.

EL NIÑO SOUTHERN OSCILLATION (ENSO). A coupled mode of ocean-atmosphere variability, centered in the tropical Pacific, and the single most important driver of interannual global climate variability, cycling with a frequency of ~2 to 7 years. Warm (El Niño) and cold (La Niña) phases of ENSO have major impacts on hydroclimate and drought in many regions, including the Maritime Continent, Australia, monsoon Asia, southern Africa, East Africa, the Amazon, and southwestern North America.

EVAPOTRANSPIRATION (ET). The flux of water from the ocean or land surface to the atmosphere, representing the phase change from liquid to gas (water vapor). Over land, ET is divided into three components: physical evaporation from the soil surface (*soil evaporation*), physical evaporation from plant surfaces (*interception losses*), and transpiration (*water taken up by the roots of plants and lost to the atmosphere through stomata*). For many continuously vegetated regions, the total ET flux is dominated by transpiration for two reasons: (1) leaves provide a larger effective surface area for ET, and (2) roots can access deeper soil moisture pools that are otherwise disconnected from the surface.

FLASH DROUGHT. A drought event, typically soil moisture based, that develops rapidly over the course of days to weeks, often with little or no advance warning.

Flash droughts most typically occur during the warm season, when temperatures and evaporative demand are high. They can be forced by both precipitation deficits and high evaporative losses, if temperatures are especially warm.

FLUX. The movement or transfer of water between reservoirs. For example, water is transferred from the atmosphere to the ocean or land surface primarily through precipitation (primarily rain and snow); this water then returns to the atmosphere from the surface via evapotranspiration. Fluxes of water in the Earth system are mediated by a variety of physical and biological processes, many of which are expected to respond significantly to climate change.

GREEN SAHARA. A period during the mid-Holocene (~9,000 to 6,000 years ago) when the West African monsoon intensified and expanded across the Sahara, shifting the entire region to a much more mesic climate compared to today. Conditions during this period were wet enough to support a large-scale expansion of savanna ecosystems and human settlements across the region. This wetting was likely initially forced by orbitally-driven increases in boreal summer insolation, with amplifying feedbacks from the land surface, vegetation, and dust aerosols. With the decline in insolation over the Holocene, and in concert with these same land surface feedbacks, the Green Sahara abruptly collapsed, shifting to a much more arid state.

GREEN WATER. Water stored in soils, recharged primarily by precipitation, and eventually lost to the atmosphere through evapotranspiration by crops and natural vegetation. Green water represents the majority of global human water use (~61 percent) but has been traditionally more difficult to proactively manage compared to blue water.

GREENHOUSE EFFECT. Heating of the Earth's surface by the atmosphere from the absorption and emission of infrared radiation by *greenhouse gases* (e.g., CO_2, CH_4, water vapor).

GROUNDWATER. Water stored underground in aquifers. Globally, groundwater represents the single largest source (~30 percent of the global total) of available freshwater. However, over many regions, pumping and withdrawal of groundwater is intense and exceeds natural recharge rates. In such areas, the depletion and eventual loss of this critical water resource are likely to negatively impact resiliency and water security.

HADLEY CIRCULATION. The thermally direct circulation cells in the atmosphere in the tropics and subtropics. Intense solar insolation in the tropics causes strong surface heating, evapotranspiration, rising motion, convergence of moisture, and high precipitation along the Intertropical Convergence Zone. Aloft, this air travels poleward before eventually subsiding and sinking ~30°N and ~30°S latitude. This subsiding air warms as it sinks, resulting in a belt of semipermanent high pressure across the subtropics, suppressing precipitation, and creating some of the largest desert regions in the world (e.g., Sahara). On the equatorward side of these high-pressure regions, the air returns to the tropics at the surface as the northeast and southeast trade winds, closing the circulation. The Hadley circulation is not fixed in time and migrates north and south over the year, following the seasonal migration of the latitude of maximum solar heating (i.e., the thermal equator).

HOLOCENE. The current interglacial period that began following the Younger Dryas event ~11,700 years ago. The long-term, millennial-scale climate evolution of the Holocene was dominated by insolation changes from orbital forcing and the retreat of the ice sheets, which also caused major shifts in regional hydroclimate (e.g., the Green Sahara). Hydroclimate events in more recent millennia (e.g., the last 2,000 years) appear more closely related to internal variability in the climate system (e.g., the megadroughts during the Medieval Climate Anomaly) rather than any specific external forcing.

HUMIDITY. A suite of variables describing the moisture state or content of the atmosphere. Humidity can refer to the saturation state of the atmosphere (e.g., saturation vapor pressure), the actual moisture content (e.g., specific humidity or the mixing ratio of water vapor), or the ratio or difference between the two (relative humidity or vapor pressure deficit).

HYDRAULIC FAILURE. A hypothesized mechanism for vegetation mortality due to drought, arising from cavitation of xylem vessels in plants. This occurs when the demand for moisture by the canopy exceeds the availability of moisture in the soil or stem of the plant, causing aspiration of the water column in the xylem.

INDIAN OCEAN DIPOLE. A zonal oscillation in sea surface temperatures across the Indian Ocean. Variations in the strength and phase of the dipole have strong impacts on drought and hydroclimate in land regions surrounding the Indian Ocean, including East Africa, India, the Maritime Continent, and Australia.

INFILTRATION. Penetration of water (typically rain or snowmelt) into the soil. Infiltration rates depend on a variety of factors, including physical soil properties, moisture already within the soil, presence or absence of vegetation, and intensity of rainfall or snowmelt. Intense precipitation rates, for example, can quickly overcome the infiltration capacity of many soils, generating significant surface runoff.

INTERDECADAL PACIFIC OSCILLATION/PACIFIC DECADAL OSCILLATION (IPO/PDO). A low-frequency (decadal) mode of climate variability in the Pacific Ocean, with spatial patterns of sea surface temperature anomalies similar to those of the El Niño Southern Oscillation (ENSO). A cold IPO/PDO phase is analogous to La Niña, and a warm IPO/PDO phase is analogous to El Niño. The hydroclimate impacts of the IPO/PDO are also similar to those of ENSO (e.g., a cold phase of the IPO/PDO favors drought across southwestern North America), and there is a tendency for impacts to be amplified when both the IPO/PDO and ENSO are in phase (i.e., when a cold IPO/PDO occurs simultaneously with a La Niña event).

IRRIGATION. Water that is applied to agricultural fields in order to grow crops. Irrigation is drawn from both natural water sources (e.g., lakes, rivers, and groundwater aquifers) and human-made water sources (e.g., artificial surface reservoirs). Rapid intensification of irrigation over the latter half of the 20th century, along with other agricultural improvements, has led to massive increases in crop production. For example, the ~20 percent of global croplands that are irrigated are responsible for ~40 percent of global agricultural production. The feasibility of expanding or even maintaining these irrigation rates into the future is uncertain, however, especially in the face of declining water resources (e.g., groundwater depletion) and the expected changes in hydroclimate with climate change.

LAND DEGRADATION/DESERTIFICATION. Declines in ecosystem structure and function and the loss of ecosystem services due to climate variability and change (e.g., the Sahel drought), direct human interventions (e.g., deforestation and overstocking of livestock), or other factors (e.g., shrub encroachment in the U.S. Southwest). Land degradation that occurs in drylands is referred to as desertification. Land degradation can manifest in a variety of ways, including changes in species composition, declines in biological productivity, increases in erosion, and declines in soil quality. Because of biological and physical feedbacks at local and regional scales, land degradation is often difficult to reverse, even if the initial cause is removed.

LA NIÑA. See *El Niño Southern Oscillation (ENSO)*.

LITTLE ICE AGE (LIA). A multicentury period (~1300 to 1800) of relatively cool conditions in the Northern Hemisphere, especially over the North Atlantic region. Even though the cause is not completely clear, there is strong evidence that it may have been started and then maintained by the combined influence of a cluster of strong volcanic eruptions and low solar forcing.

MEDIEVAL CLIMATE ANOMALY (MCA). A multicentury period (~800 to 1300 CE) of relatively warm conditions across the Northern Hemisphere. The MCA stands out from a hydroclimate perspective because this was a period of extensive and persistent multidecadal drought across much of North America, especially California, the southwestern region, and the central plains.

MEDITERRANEAN CLIMATE. A climate characterized by wet, cool winters and dry, warm summers, typically located along the west coasts of continents in the midlatitudes. Proximity to the ocean moderates the seasonal amplitude of temperature, whereas the precipitation seasonality is controlled by the seasonal migration of the semipermanent high-pressure centers sitting over the oceans. Mediterranean regions are especially prone to drought and fire because the dry season corresponds to the warmest time of year (summer) when atmospheric demand for moisture is at its annual maximum.

MEGADROUGHT. A term typically used to describe a persistent, usually multidecadal drought event, especially those that occurred over areas of North America (California, the southwestern region, and the central plains) during the Medieval Climate Anomaly. The causes of these megadroughts are not well established, but there is strong evidence that exceptional ocean conditions (a cold eastern tropical Pacific and warm tropical Atlantic) were a major driver, especially over southwestern North America. Additional hypotheses have highlighted the influence of internal atmospheric variability and land surface feedbacks, especially for megadroughts that occurred over the central plains.

MOISTURE DIVERGENCE/CONVERGENCE. The horizontal transport of water in the atmosphere. *Divergence* refers to a net export of moisture out of a specific location, whereas *convergence* refers to a net import of moisture. In the long-term average (i.e., when changes in storage in the atmosphere at a given location average out to zero), moisture divergence must be balanced by evapotranspiration minus precipitation.

MONSOON CLIMATE. A climate typically associated with strong seasonal wind reversals and characterized by hot, wet summers and extended dry winters. Areas of monsoon climate are located primarily in the tropics and subtropics (e.g., West Africa and India).

NORTH ATLANTIC OSCILLATION (NAO). A mode of climate variability in the North Atlantic, modulated by the pressure differential between the Icelandic Low and Azores High. Impacts of the NAO on hydroclimate are especially strong over Europe during winter and spring. During a positive phase of the NAO (characterized by a strong pressure gradient between the two centers of action), storm tracks are shifted northward, and northern Europe is warm and wet while drought conditions occur over much of the Mediterranean. This pattern of climate anomalies in Europe is reversed when the NAO is in a negative phase (characterized by a weak pressure gradient).

PALEOCLIMATOLOGY. The study of past climate variability and change, especially for time periods prior to the availability of high-quality instrumental records. Paleoclimate studies use various geological and biological proxies (e.g., tree rings, ice cores, and lake sediments) to infer information about the climate system in the past. Studies of hydroclimate in the paleoclimate record have identified major drought and pluvial events, highlighting a much larger range of variability in the climate system compared to the last 150 years of much more limited historical observations.

PLUVIAL. An anomalous moisture surplus, effectively the opposite of a drought. Pluvials are not the same as floods or intense precipitation events. Rather, the term refers to extended periods (e.g., months to years) of above-average moisture availability.

POTENTIAL EVAPOTRANSPIRATION (PET). Also referred to as reference evapotranspiration or evaporative demand, PET is the theoretical maximum evapotranspiration that could occur given unlimited water availability at the surface. Potential evapotranspiration depends on a variety of factors, including temperature, humidity, wind speed, radiation, and resistance from the land surface. In most cases, actual evapotranspiration rarely reaches the full potential evapotranspiration rate. However, PET is still a useful quantity in a variety of hydroclimate applications, as it reflects the atmospheric demand for moisture.

PRECIPITATION. Liquid or solid water particles that fall from the atmosphere to the land or ocean surface. From the perspective of hydroclimate and drought, most discussions of precipitation are centered on *rainfall* (liquid precipitation) and *snowfall* (solid precipitation made up of ice crystals).

RESERVOIR. Any physical subdivision that contains an amount of available water we can measure and that can exchange water with other reservoirs through fluxes (e.g., precipitation, evapotranspiration, and runoff). At local and regional scales, examples include natural lakes and rivers, human-created impoundments, and aquifers. At the scale of the global water cycle, the atmosphere and oceans can each be viewed as singular reservoirs. In situations where inflows are equal to outflows, the system is considered to be in equilibrium (i.e., the net amount of water contained in the reservoir will not change over time).

RESIDENCE TIME. The average time a particle spends in a reservoir, equivalent to the time it takes for the entire volume of a reservoir to turn over (i.e., be completely replaced).

RUNOFF. In the surface moisture budget, horizontal flows of water representing the moisture available after accounting for surface inputs from precipitation and losses due to evapotranspiration. Runoff occurs at the surface and belowground and is the critical source of recharge for many lakes, rivers, and aquifers. Runoff is sensitive to a variety of processes, including soil properties, precipitation intensity, vegetation, and soil and snow water storage. All these processes affect the *runoff efficiency*, which is the amount of runoff generated per unit of precipitation.

SNOW DROUGHT. An anomalous deficit in the size or water content of the snowpack. Snow droughts are caused by deficits in total precipitation but can also occur because of anomalously warm temperatures. In the latter case, often termed a *wet* snow drought, this occurs because warmer temperatures increase the fraction of precipitation falling as rain rather than snow and reduce the snowpack size by increasing losses from sublimation and melting.

SNOW WATER EQUIVALENT (SWE). The amount of actual water contained within a unit of snowpack if all the snow was melted. From a hydroclimate perspective, SWE is more useful than snow depth because the density of snow can change depending on temperature and other processes.

STOMATAL CONDUCTANCE. The rate of exchange of gases between the leaves of plants and the atmosphere, usually in reference to carbon dioxide and water vapor. For water vapor, conductance can depend on a variety of factors, including the humidity, carbon dioxide concentrations in the atmosphere, and leaf water content. In response to drought, many plants will close their stomata, decreasing their stomatal conductance to help conserve water.

TELECONNECTION. The physical linkages between weather and climate events across geographically distant regions. A good example is the El Niño Southern Oscillation, a phenomenon centered in the tropical Pacific that has far-reaching implications for drought and hydroclimate variability on nearly every continent.

THROUGHFALL & STEMFLOW. Precipitation that reaches the ground surface either by passing directly through the canopy without being intercepted (*throughfall*) or that flows down or drips off plants surfaces (*stemflow*). Together, these represent the fraction of precipitation not intercepted by the canopy that is available to infiltrate into the soils or generate runoff.

VAPOR PRESSURE DEFICIT (VPD). The difference between saturated and actual vapor pressure in the atmosphere. VPD is a useful indicator of the atmospheric demand for moisture and overall aridity.

WALKER CIRCULATION. Zonal (east-west) circulation in the atmosphere over the tropical Pacific, characterized by rising motion in the west over the Maritime Continent and Indo-Pacific Warm Pool, eastward flow aloft, subsidence in the east near the west coast of South America, and westward flow at the surface. The strength of the Walker circulation is strongly controlled by variability in the El Niño Southern Oscillation.

WATER SCARCITY. A situation where available water resources are insufficient to meet population demands in a region.

WITHDRAWAL AND CONSUMPTION. The two categories of human water use. *Withdrawal* refers to the total amount of water taken from a source, some of which may return to the source to eventually be withdrawn again. *Consumption* is water that is used but is no longer available for reuse.

References

Abatzoglou, J. T., & Williams, A. P. (2016). Impact of anthropogenic climate change on wildfire across western US forests. *Proceedings of the National Academy of Sciences*, 113 (42), 11770–11775.

Adams, H. D., Zeppel, M. J., Anderegg, W. R., Hartmann, H., Landhäusser, S. M., Tissue, D. T., ... & Anderegg, L. D. (2017). A multi-species synthesis of physiological mechanisms in drought-induced tree mortality. *Nature Ecology and Evolution*, 1 (9), 1285.

Afifi, T. (2011). Economic or environmental migration? The push factors in Niger. *International Migration*, 49, e95–e124.

AghaKouchak, A. (2015). Recognize anthropogenic drought. *Nature*, 524 (7566), 409.

Allen, C. D., & Breshears, D. D. (1998). Drought-induced shift of a forest–woodland ecotone: Rapid landscape response to climate variation. *Proceedings of the National Academy of Sciences*, 95 (25), 14839–14842.

Alter, R. E., Fan, Y., Lintner, B. R., & Weaver, C. P. (2015). Observational evidence that Great Plains irrigation has enhanced summer precipitation intensity and totals in the midwestern United States. *Journal of Hydrometeorology*, 16 (4), 1717–1735.

Alter, R. E., Im, E. S., & Eltahir, E. A. (2015). Rainfall consistently enhanced around the Gezira Scheme in East Africa due to irrigation. *Nature Geoscience*, 8 (10), 763.

Anderegg, W. R., Schwalm, C., Biondi, F., Camarero, J. J., Koch, G., Litvak, M., ... & Wolf, A. (2015). Pervasive drought legacies in forest ecosystems and their implications for carbon cycle models. *Science*, 349 (6247), 528–532.

Anselmetti, F. S., Hodell, D. A., Ariztegui, D., Brenner, M., & Rosenmeier, M. F. (2007). Quantification of soil erosion rates related to ancient Maya deforestation. *Geology*, 35 (10), 915–918.

Asner, G. P., Scurlock, J. M., & Hicke, J. A. (2003). Global synthesis of leaf area index observations: Implications for ecological and remote sensing studies. *Global Ecology and Biogeography*, 12 (3), 191–205.

Aubee, E., & Hussein, K. (2002). Emergency relief, crop diversification and institution building: The case of sesame in the Gambia. *Disasters*, 26 (4), 368–379.

Ault, T. R., St. George, S., Smerdon, J. E., Coats, S., Mankin, J. S., Carrillo, C. M., . . . & Stevenson, S. (2018). A robust null hypothesis for the potential causes of megadrought in western North America. *Journal of Climate*, 31 (1), 3–24.

Bai, Z. G., Dent, D. L., Olsson, L., & Schaepman, M. E. (2008). Proxy global assessment of land degradation. *Soil Use and Management*, 24 (3), 223–234.

Barlow, P. M., & Reichard, E. G. (2010). Saltwater intrusion in coastal regions of North America. *Hydrogeology Journal*, 18 (1), 247–260.

Barnett, T. P., Pierce, D. W., AchutaRao, K. M., Gleckler, P. J., Santer, B. D., Gregory, J. M., & Washington, W. M. (2005). Penetration of human-induced warming into the world's oceans. *Science*, 309 (5732), 284–287.

Barnston, A. G., & Schickedanz, P. T. (1984). The effect of irrigation on warm season precipitation in the southern Great Plains. *Journal of Climate and Applied Meteorology*, 23 (6), 865–888.

Bates, B., Kundzewicz, Z. W., Wu, S., & Palutikof, J. P. (Eds.) (2008). *Climate Change and Water* (Technical Paper of the Intergovernmental Panel on Climate Change). Geneva: Intergovernmental Panel on Climate Change Secretariat.

Beer, J. (2000). Long-term indirect indices of solar variability. *Space Science Reviews*, 94 (1–2), 53–66.

Behling, H. (1995). Investigations into the Late Pleistocene and Holocene history of vegetation and climate in Santa Catarina (S Brazil). *Vegetation History and Archaeo-botany*, 4 (3), 127–152.

Behling, H., & Pillar, V. D. (2007). Late Quaternary vegetation, biodiversity and fire dynamics on the southern Brazilian highland and their implication for conservation and management of modern Araucaria forest and grassland ecosystems. *Philosophical Transactions of the Royal Society B: Biological Sciences*, 362 (1478), 243–251.

Belmecheri, S., Babst, F., Wahl, E. R., Stahle, D. W., & Trouet, V. (2016). Multi-century evaluation of Sierra Nevada snowpack. *Nature Climate Change*, 6 (1), 2.

Benson, L. V., Berry, M. S., Jolie, E. A., Spangler, J. D., Stahle, D. W., & Hattori, E. M. (2007). Possible impacts of early-11th-, middle-12th-, and late-13th-century droughts on western Native Americans and the Mississippian Cahokians. *Quaternary Science Reviews*, 26 (3–4), 336–350.

Benson, L. V., Pauketat, T. R., & Cook, E. R. (2009). Cahokia's boom and bust in the context of climate change. *American Antiquity*, 74 (3), 467–483.

Benson, L., Petersen, K., & Stein, J. (2007). Anasazi (pre-Columbian Native-American) migrations during the middle-12th and late-13th centuries—Were they drought induced? *Climatic Change*, 83 (1–2), 187–213.

Berg, A., Findell, K., Lintner, B., Giannini, A., Seneviratne, S. I., Van den Hurk, B., . . . & Cheruy, F. (2016). Land–atmosphere feedbacks amplify aridity increase over land under global warming. *Nature Climate Change*, 6 (9), 869.

Berg, A., Sheffield, J., & Milly, P. C. (2017). Divergent surface and total soil moisture projections under global warming. *Geophysical Research Letters*, 44 (1), 236–244.

Berg, N., & Hall, A. (2015). Increased interannual precipitation extremes over California under climate change. *Journal of Climate, 28* (16), 6324–6334.

Berg, N., & Hall, A. (2017). Anthropogenic warming impacts on California snowpack during drought. *Geophysical Research Letters, 44* (5), 2511–2518.

Berkelhammer, M., Sinha, A., Stott, L., Cheng, H., Pausata, F. S., & Yoshimura, K. (2012). An abrupt shift in the Indian monsoon 4000 years ago. In L. Giosan, D. Q. Fuller, K. Nicoll, R. K. Flad, & P. D. Clift (Eds.), *Climates, landscapes, and civilizations* (pp. 75–88). Washington, DC: American Geophysical Union.

Bernus, E. (1990). Dates, dromedaries, and drought: Diversification in Tuareg pastoral systems. In J. G. Galaty & D. L. Johnson (Eds.), *The world of pastoralism: Herding systems in comparative perspective* (pp. 149–176). New York: Belhaven.

Biasutti, M. (2013). Forced Sahel rainfall trends in the CMIP5 archive. *Journal of Geophysical Research: Atmospheres, 118* (4), 1613–1623.

Biasutti, M., & Giannini, A. (2006). Robust Sahel drying in response to late 20th century forcings. *Geophysical Research Letters, 33* (11).

Bilotta, G. S., Brazier, R. E., & Haygarth, P. M. (2007). The impacts of grazing animals on the quality of soils, vegetation, and surface waters in intensively managed grasslands. *Advances in Agronomy, 94*, 237–280.

Bindoff, N. L., Stott, P. A., AchutaRao, K. M., Allen, M. R., Gillett, N., Gutzler, D., . . . & Zhang, X. (2013). Detection and attribution of climate change: From global to regional. In T. F. Stocker, D. Qin, G.-K. Plattner, M. Tignor, S. K. Allen, J. Boschung, . . . & P. M. Midgley (Eds.), *Climate change 2013: The physical science basis: Working Group I contribution to the Fifth Assessment Report of the Intergovernmental Panel on Climate Change.* Cambridge, UK: Cambridge University Press.

Bonan, G. B. (2008). Forests and climate change: Forcings, feedbacks, and the climate benefits of forests. *Science, 320* (5882), 1444–1449.

Bonfils, C., & Lobell, D. (2007). Empirical evidence for a recent slowdown in irrigation-induced cooling. *Proceedings of the National Academy of Sciences, 104* (34), 13582–13587.

Booth, B. B., Dunstone, N. J., Halloran, P. R., Andrews, T., & Bellouin, N. (2012). Aerosols implicated as a prime driver of twentieth-century North Atlantic climate variability. *Nature, 484* (7393), 228.

Bot, A., Nachtergaele, F., & Young, A. (2000). *Land resource potential and constraints at regional and country levels* (World Soil Resources Report No. 90). Rome: Food and Agriculture Organization of the United Nations.

Boucher, O., Myhre, G., & Myhre, A. (2004). Direct human influence of irrigation on atmospheric water vapour and climate. *Climate Dynamics, 22* (6–7), 597–603.

Braconnot, P., Harrison, S. P., Joussaume, S., Hewitt, C. D., Kitoch, A., Kutzbach, J. E., . . . & Weber, S. L. (2004). Evaluation of PMIP coupled ocean-atmosphere simulations of the mid-Holocene. In R. W. Battarbee, F. Gasse, & C. E. Stickley (Eds.), *Past climate variability through Europe and Africa* (pp. 515–533). Dordrecht, Netherlands: Springer.

Brandt, M., Mbow, C., Diouf, A. A., Verger, A., Samimi, C., & Fensholt, R. (2015). Ground-and satellite-based evidence of the biophysical mechanisms behind the greening Sahel. *Global Change Biology, 21* (4), 1610–1620.

Brower, C., Prins, K., & Heibloem, M. (1989). *Irrigation water management: Irrigation scheduling* (Irrigation Water Management Training Manual No. 4). Rome: Food and Agriculture Organization of the United Nations. Retrieved from http://www.fao .org/docrep/t7202e/t7202e00.htm.

Buckley, B. M., Anchukaitis, K. J., Penny, D., Fletcher, R., Cook, E. R., Sano, M., . . . & Hong, T. M. (2010). Climate as a contributing factor in the demise of Angkor, Cambodia. *Proceedings of the National Academy of Sciences*, 107 (15), 6748–6752.

Cai, W., Van Rensch, P., Cowan, T., & Sullivan, A. (2010). Asymmetry in ENSO teleconnection with regional rainfall, its multidecadal variability, and impact. *Journal of Climate*, 23 (18), 4944–4955.

Cai, X., Zhang, X., & Wang, D. (2011). Land availability for biofuel production. *Environmental Science and Technology*, 45 (1), 334–339.

Campbell, J. E., Lobell, D. B., Genova, R. C., & Field, C. B. (2008). The global potential of bioenergy on abandoned agriculture lands. *Environmental Science and Technology*, 42 (15), 5791–5794.

Canadell, J., Jackson, R. B., Ehleringer, J. B., Mooney, H. A., Sala, O. E., & Schulze, E. D. (1996). Maximum rooting depth of vegetation types at the global scale. *Oecologia*, 108 (4), 583–595.

Cárdenas, M. F., Tobón, C., & Buytaert, W. (2017). Contribution of occult precipitation to the water balance of páramo ecosystems in the Colombian Andes. *Hydrological Processes*, 31 (24), 4440–4449.

Carré, M., Sachs, J. P., Purca, S., Schauer, A. J., Braconnot, P., Falcón, R. A., . . . & Lavallée, D. (2014). Holocene history of ENSO variance and asymmetry in the eastern tropical Pacific. *Science*, 345 (6200), 1045–1048.

Chang, C. Y., Chiang, J. C. H., Wehner, M. F., Friedman, A. R., & Ruedy, R. (2011). Sulfate aerosol control of tropical Atlantic climate over the twentieth century. *Journal of Climate*, 24 (10), 2540–2555.

Charney, J. G. (1975). Dynamics of deserts and drought in the Sahel. *Quarterly Journal of the Royal Meteorological Society*, 101 (428), 193–202.

Chepil, W. S., Siddoway, F. H., & Armbrust, D. (1963). Climatic index of wind erosion conditions in the Great Plains. *Soil Science Society of America Journal*, 27 (4), 449–452.

Christensen, N. S., Wood, A. W., Voisin, N., Lettenmaier, D. P., & Palmer, R. N. (2004). The effects of climate change on the hydrology and water resources of the Colorado River Basin. *Climatic Change*, 62 (1–3), 337–363.

Claussen, M., Kubatzki, C., Brovkin, V., Ganopolski, A., Hoelzmann, P., & Pachur, H. J. (1999). Simulation of an abrupt change in Saharan vegetation in the mid-Holocene. *Geophysical Research Letters*, 26 (14), 2037–2040.

Coats, S., Smerdon, J. E., Cook, B. I., & Seager, R. (2015). Are simulated megadroughts in the North American southwest forced? *Journal of Climate*, 28 (1), 124–142.

Coats, S., Smerdon, J. E., Cook, B. I., Seager, R., Cook, E. R., & Anchukaitis, K. J. (2016). Internal ocean-atmosphere variability drives megadroughts in western North America. *Geophysical Research Letters*, 43 (18), 9886–9894.

Coats, S., Smerdon, J. E., Karnauskas, K. B., & Seager, R. (2016). The improbable but unexceptional occurrence of megadrought clustering in the American West during the Medieval Climate Anomaly. *Environmental Research Letters*, 11 (7), 074025.

Coats, S., & Smerdon, J. E. (2017). Climate variability: The Atlantic's internal drum beat. *Nature Geoscience*, 10 (7), 470.

Cobb, K. M., Westphal, N., Sayani, H. R., Watson, J. T., Di Lorenzo, E., Cheng, H., . . . & Charles, C. D. (2013). Highly variable El Niño–Southern Oscillation throughout the Holocene. *Science*, 339 (6115), 67–70.

Collins, M., Knutti, R., Arblaster, J., Dufresne, J.-L., Fichefet, T., Friedlingstein, P., . . . & Wehner, M. (2013). Long-term climate change: Projections, commitments and irreversibility. In T. F. Stocker, D. Qin, G.-K. Plattner, M. Tignor, S. K. Allen, J. Boschung, . . . & P. M. Midgley (Eds.), *Climate change 2013: The physical science basis: Working Group I contribution to the Fifth Assessment Report of the Intergovernmental Panel on Climate Change*. Cambridge, UK: Cambridge University Press.

Cook, B. I., Anchukaitis, K. J., Kaplan, J. O., Puma, M. J., Kelley, M., & Gueyffier, D. (2012). Pre-Columbian deforestation as an amplifier of drought in Mesoamerica. *Geophysical Research Letters*, 39 (16).

Cook, B. I., Anchukaitis, K. J., Touchan, R., Meko, D. M., & Cook, E. R. (2016). Spatio-temporal drought variability in the Mediterranean over the last 900 years. *Journal of Geophysical Research: Atmospheres*, 121 (5), 2060–2074.

Cook, B. I., Ault, T. R., & Smerdon, J. E. (2015). Unprecedented 21st century drought risk in the American Southwest and central plains. *Science Advances*, 1 (1), e1400082.

Cook, B. I., Cook, E. R., Smerdon, J. E., Seager, R., Williams, A. P., Coats, S., . . . & Díaz, J. V. (2016). North American megadroughts in the Common Era: Reconstructions and simulations. *Wiley Interdisciplinary Reviews: Climate Change*, 7 (3), 411–432.

Cook, B. I., Miller, R. L., & Seager, R. (2009). Amplification of the North American "Dust Bowl" drought through human-induced land degradation. *Proceedings of the National Academy of Sciences*, 106 (13), 4997–5001.

Cook, B. I., Seager, R., & Miller, R. L. (2011). On the causes and dynamics of the early twentieth-century North American pluvial. *Journal of Climate*, 24 (19), 5043–5060.

Cook, B. I., Seager, R., Miller, R. L., & Mason, J. A. (2013). Intensification of North American megadroughts through surface and dust aerosol forcing. *Journal of Climate*, 26 (13), 4414–4430.

Cook, B. I., Seager, R., & Smerdon, J. E. (2014). The worst North American drought year of the last millennium: 1934. *Geophysical Research Letters*, 41 (20), 7298–7305.

Cook, B. I., Shukla, S. P., Puma, M. J., & Nazarenko, L. S. (2015). Irrigation as an historical climate forcing. *Climate Dynamics*, 44 (5–6), 1715–1730.

Cook, B. I., Smerdon, J. E., Seager, R., & Coats, S. (2014). Global warming and 21st century drying. *Climate Dynamics*, 43 (9–10), 2607–2627.

Cook, E. R., Anchukaitis, K. J., Buckley, B. M., D'Arrigo, R. D., Jacoby, G. C., & Wright, W. E. (2010). Asian monsoon failure and megadrought during the last millennium. *Science*, 328 (5977), 486–489.

Cook, E. R., Seager, R., Heim, R. R., Vose, R. S., Herweijer, C., & Woodhouse, C. (2010). Megadroughts in North America: Placing IPCC projections of hydroclimatic change in a long-term palaeoclimate context. *Journal of Quaternary Science, 25* (1), 48–61.

Cook, K. H., & Vizy, E. K. (2006). Coupled model simulations of the West African monsoon system: Twentieth- and twenty-first-century simulations. *Journal of Climate, 19* (15), 3681–3703.

Dahl, J. (1940). Progress and development of the prairie states forestry project. *Journal of Forestry, 38* (4), 301–306.

Dardel, C., Kergoat, L., Hiernaux, P., Mougin, E., Grippa, M., & Tucker, C. J. (2014). Re-greening Sahel: 30 years of remote sensing data and field observations (Mali, Niger). *Remote Sensing of Environment, 140,* 350–364.

Davis, M. (2002). *Late Victorian holocausts: El Niño famines and the making of the third world.* London: Verso Books.

Dawson, T. E. (1998). Fog in the California redwood forest: Ecosystem inputs and use by plants. *Oecologia, 117* (4), 476–485.

DeAngelis, A., Dominguez, F., Fan, Y., Robock, A., Kustu, M. D., & Robinson, D. (2010). Evidence of enhanced precipitation due to irrigation over the Great Plains of the United States. *Journal of Geophysical Research: Atmospheres, 115* (D15).

de Bruijn, M. (1997). The hearthhold in pastoral Fulbe society, central Mali: Social relations, milk and drought. *Africa, 67* (4), 625–651.

Delworth, T. L., Zeng, F., Rosati, A., Vecchi, G. A., & Wittenberg, A. T. (2015). A link between the hiatus in global warming and North American drought. *Journal of Climate, 28* (9), 3834–3845.

deMenocal, P. B. (2001). Cultural responses to climate change during the late Holocene. *Science, 292* (5517), 667–673.

deMenocal, P. B., & Tierney, J. E. (2012). Green Sahara: African humid periods paced by Earth's orbital changes. *Nature Education Knowledge, 3* (10), 12.

Diffenbaugh, N. S., Swain, D. L., & Touma, D. (2015). Anthropogenic warming has increased drought risk in California. *Proceedings of the National Academy of Sciences, 112* (13), 3931–3936.

Dilley, M., Chen, R. S., Deichmann, U., Lerner-Lam, A. L., & Arnold, M. (2005). *Natural disaster hotspots: A global risk analysis* (Disaster Risk Management Series No. 5). Washington, DC: World Bank.

D'Odorico, P., Bhattachan, A., Davis, K. F., Ravi, S., & Runyan, C. W. (2013). Global desertification: Drivers and feedbacks. *Advances in Water Resources, 51,* 326–344.

Donarummo, J., Ram, M., & Stoermer, E. F. (2003). Possible deposit of soil dust from the 1930's US dust bowl identified in Greenland ice. *Geophysical Research Letters, 30* (6).

Dong, B., & Dai, A. (2015). The influence of the interdecadal Pacific oscillation on temperature and precipitation over the globe. *Climate Dynamics, 45* (9–10), 2667–2681.

Donohue, R. J., Roderick, M. L., McVicar, T. R., & Farquhar, G. D. (2013). Impact of CO_2 fertilization on maximum foliage cover across the globe's warm, arid environments. *Geophysical Research Letters, 40* (12), 3031–3035.

Douglass, A. E. (1929). The secret of the Southwest solved by talkative tree rings. *National Geographic*, 56, 736–770.

Douglass, A. E. (1935). *Dating Pueblo Bonito and other ruins of the Southwest*. Washington, DC: National Geographic Society.

Douglas, P. M., Demarest, A. A., Brenner, M., & Canuto, M. A. (2016). Impacts of climate change on the collapse of lowland Maya civilization. *Annual Review of Earth and Planetary Sciences*, 44, 613–645.

Ellis, E. C., Kaplan, J. O., Fuller, D. Q., Vavrus, S., Goldewijk, K. K., & Verburg, P. H. (2013). Used planet: A global history. *Proceedings of the National Academy of Sciences*, 110 (20), 7978–7985.

Eltahir, E. A., & Bras, R. L. (1994). Precipitation recycling in the Amazon basin. *Quarterly Journal of the Royal Meteorological Society*, 120 (518), 861–880.

Eltahir, E. A., & Bras, R. L. (1996). Precipitation recycling. *Reviews of Geophysics*, 34 (3), 367–378.

Emile-Geay, J., Cobb, K. M., Carré, M., Braconnot, P., Leloup, J., Zhou, Y., . . . & Driscoll, R. (2016). Links between tropical Pacific seasonal, interannual and orbital variability during the Holocene. *Nature Geoscience*, 9 (2), 168.

Engelstaedter, S., Tegen, I., & Washington, R. (2006). North African dust emissions and transport. *Earth-Science Reviews*, 79 (1–2), 73–100.

Epule, E. T., Peng, C., Lepage, L., & Chen, Z. (2014). The causes, effects and challenges of Sahelian droughts: A critical review. *Regional Environmental Change*, 14 (1), 145–156.

Erban, L. E., Gorelick, S. M., & Zebker, H. A. (2014). Groundwater extraction, land subsidence, and sea-level rise in the Mekong Delta, Vietnam. *Environmental Research Letters*, 9 (8), 084010.

Famiglietti, J. S. (2014). The global groundwater crisis. *Nature Climate Change*, 4 (11), 945.

FAO (Food and Agriculture Organization of the United Nations). (2016). AQUASTAT database. http://www.fao.org/nr/water/aquastat/main/index.stm.

Faurès, J. M., Hoogeveen, J., & Bruinsma, J. (2002). *The FAO irrigated area forecast for 2030*. Rome, Italy: Food and Agriculture Organization of the United Nations.

Ferguson, G., & Gleeson, T. (2012). Vulnerability of coastal aquifers to groundwater use and climate change. *Nature Climate Change*, 2 (5), 342.

Ferguson, I. M., & Maxwell, R. M. (2010). Role of groundwater in watershed response and land surface feedbacks under climate change. *Water Resources Research*, 46 (10).

Finné, M., Holmgren, K., Sundqvist, H. S., Weiberg, E., & Lindblom, M. (2011). Climate in the eastern Mediterranean, and adjacent regions, during the past 6000 years–A review. *Journal of Archaeological Science*, 38 (12), 3153–3173.

Fischer, E. M., Seneviratne, S. I., Lüthi, D., & Schär, C. (2007). Contribution of land-atmosphere coupling to recent European summer heat waves. *Geophysical Research Letters*, 34 (6).

Forman, S. L., Oglesby, R., & Webb, R. S. (2001). Temporal and spatial patterns of Holocene dune activity on the Great Plains of North America: Megadroughts and climate links. *Global and Planetary Change*, 29 (1–2), 1–29.

Freydank, K., & Siebert, S. (2008). *Towards mapping the extent of irrigation in the last century: Time series of irrigated area per country*. Frankfurt am Main: University of Frankfurt (Main).

Fritz, S. C., Metcalfe, S. E., & Dean, W. (2001). Holocene climate patterns in the Americas inferred from paleolimnological records. In V. Markgraf (Ed.), *Interhemispheric climate linkages* (pp. 241–263). San Diego, CA: Academic Press.

Galloway, D. L., & Burbey, T. J. (2011). Regional land subsidence accompanying ground-water extraction. *Hydrogeology Journal*, 19 (8), 1459–1486.

García-Herrera, R., Díaz, J., Trigo, R. M., Luterbacher, J., & Fischer, E. M. (2010). A review of the European summer heat wave of 2003. *Critical Reviews in Environmental Science and Technology*, 40 (4), 267–306.

Gautier, D., Denis, D., & Locatelli, B. (2016). Impacts of drought and responses of rural populations in West Africa: A systematic review. *Wiley Interdisciplinary Reviews: Climate Change*, 7 (5), 666–681.

Giannini, A. (2016). 40 years of climate modeling: The causes of late-20th century drought in the Sahel. In R. Behnke & M. Mortimore (Eds.), *The end of desertification?* (pp. 265–291). Berlin: Springer.

Giannini, A., Biasutti, M., & Verstraete, M. M. (2008). A climate model-based review of drought in the Sahel: Desertification, the re-greening and climate change. *Global and Planetary Change*, 64 (3–4), 119–128.

Giannini, A., Saravanan, R., & Chang, P. (2003). Oceanic forcing of Sahel rainfall on interannual to interdecadal time scales. *Science*, 302 (5647), 1027–1030.

Gibbs, H. K., & Salmon, J. M. (2015). Mapping the world's degraded lands. *Applied Geography*, 57, 12–21.

Gillette, D. A., & Hanson, K. J. (1989). Spatial and temporal variability of dust production caused by wind erosion in the United States. *Journal of Geophysical Research: Atmospheres*, 94 (D2), 2197–2206.

Ginoux, P., Prospero, J. M., Gill, T. E., Hsu, N. C., & Zhao, M. (2012). Global-scale attribution of anthropogenic and natural dust sources and their emission rates based on MODIS Deep Blue aerosol products. *Reviews of Geophysics*, 50 (3).

Gleeson, T., Befus, K. M., Jasechko, S., Luijendijk, E., & Cardenas, M. B. (2016). The global volume and distribution of modern groundwater. *Nature Geoscience*, 9 (2), 161–167.

Gleick, P. H. (2014). Water, drought, climate change, and conflict in Syria. *Weather, Climate, and Society*, 6 (3), 331–340.

Gleick, P. H. (2017). *Impacts of California's five-year (2012–2016) drought on hydroelectricity generation*. Oakland, CA: Pacific Institute.

Goin, D. E., Rudolph, K. E., & Ahern, J. (2017). Impact of drought on crime in California: A synthetic control approach. *PLoS One*, 12 (10), e0185629.

Gonzalez, P. (2001). Desertification and a shift of forest species in the West African Sahel. *Climate Research*, 17, 217–228.

Govaerts, Y., & Lattanzio, A. (2008). Estimation of surface albedo increase during the eighties Sahel drought from Meteosat observations. *Global and Planetary Change*, 64, 139–145.

Greve, P., Orlowsky, B., Mueller, B., Sheffield, J., Reichstein, M., & Seneviratne, S. I. (2014). Global assessment of trends in wetting and drying over land. *Nature Geoscience*, 7 (10), 716.

Griffin, D., & Anchukaitis, K. J. (2014). How unusual is the 2012–2014 California drought? *Geophysical Research Letters*, 41 (24), 9017–9023.

Grosjean, M., Cartajena, I., Geyh, M. A., & Núñez, L. (2003). From proxy data to paleoclimate interpretation: The mid-Holocene paradox of the Atacama Desert, northern Chile. *Palaeogeography, Palaeoclimatology, Palaeoecology*, 194 (1–3), 247–258.

Guimberteau, M., Laval, K., Perrier, A., & Polcher, J. (2012). Global effect of irrigation and its impact on the onset of the Indian summer monsoon. *Climate Dynamics*, 39 (6), 1329–1348.

Han, S., & Yang, Z. (2013). Cooling effect of agricultural irrigation over Xinjiang, Northwest China from 1959 to 2006. *Environmental Research Letters*, 8 (2), 024039.

Hansen, Z. K., & Libecap, G. D. (2004). Small farms, externalities, and the Dust Bowl of the 1930s. *Journal of Political Economy*, 112 (3), 665–694.

Hanson, P. R., Arbogast, A. F., Johnson, W. C., Joeckel, R. M., & Young, A. R. (2010). Megadroughts and late Holocene dune activation at the eastern margin of the Great Plains, north-central Kansas, USA. *Aeolian Research*, 1 (3–4), 101–110.

Hardin, E., AghaKouchak, A., Qomi, M. J. A., Madani, K., Tarroja, B., Zhou, Y., . . . & Samuelsen, S. (2017). California drought increases CO_2 footprint of energy. *Sustainable Cities and Society*, 28, 450–452.

Harpold, A., Dettinger, M., & Rajagopal, S. (2017). Defining snow drought and why it matters. *EOS: Earth & Space Science News*, 98.

Hart, S. J., Schoennagel, T., Veblen, T. T., & Chapman, T. B. (2015). Area burned in the western United States is unaffected by recent mountain pine beetle outbreaks. *Proceedings of the National Academy of Sciences*, 112 (14), 4375–4380.

Hastenrath, S., & Lamb, P. J. (1977). Some aspects of circulation and climate over the easternequatorial Atlantic. *Monthly Weather Review*, 105, 1019–1023.

He, M., Russo, M., & Anderson, M. (2017). Hydroclimatic characteristics of the 2012–2015 California drought from an operational perspective. *Climate*, 5 (1), 5.

Helms, D. (2009). Hugh Hammond Bennett and the creation of the soil erosion service. *Journal of Soil and Water Conservation*, 64 (2), 68A–74A.

Herrmann, S. M., Anyamba, A., & Tucker, C. J. (2005). Recent trends in vegetation dynamics in the African Sahel and their relationship to climate. *Global Environmental Change*, 15 (4), 394–404.

Hiernaux, P., Diarra, L., Trichon, V., Mougin, E., Soumaguel, N., & Baup, F. (2009). Woody plant population dynamics in response to climate changes from 1984 to 2006 in Sahel (Gourma, Mali). *Journal of Hydrology*, 375 (1–2), 103–113.

Hilker, T., Natsagdorj, E., Waring, R. H., Lyapustin, A., & Wang, Y. (2014). Satellite observed widespread decline in Mongolian grasslands largely due to overgrazing. *Global Change Biology*, 20 (2), 418–428.

Hodell, D. A., Brenner, M., Curtis, J. H., & Guilderson, T. (2001). Solar forcing of drought frequency in the Maya lowlands. *Science*, 292 (5520), 1367–1370.

Hoerling, M., Eischeid, J., Kumar, A., Leung, R., Mariotti, A., Mo, K., . . . & Seager, R. (2014). Causes and predictability of the 2012 Great Plains drought. *Bulletin of the American Meteorological Society*, 95 (2), 269–282.

Hoerling, M., Eischeid, J., Perlwitz, J., Quan, X., Zhang, T., & Pegion, P. (2012). On the increased frequency of Mediterranean drought. *Journal of Climate*, 25 (6), 2146–2161.

Hoerling, M., Quan, X. W., & Eischeid, J. (2009). Distinct causes for two principal US droughts of the 20th century. *Geophysical Research Letters*, 36 (19).

Hornbeck, R. (2012). The enduring impact of the American Dust Bowl: Short- and long-run adjustments to environmental catastrophe. *American Economic Review*, 102 (4), 1477–1507.

Howell, T. A. (2003). Irrigation: Efficiency. In B. A. Stewart & T. A. Howell (Eds.), *Encyclopedia of water science* (pp. 467–472). New York: Marcel Dekker.

Howitt, R., Medellín-Azuara, J., MacEwan, D., Lund, J. R., & Sumner, D. (2015). *Economic analysis of the 2015 drought for California agriculture.* Davis: University of California–Davis, Center for Watershed Sciences.

Huang, J., Yu, H., Guan, X., Wang, G., & Guo, R. (2016). Accelerated dryland expansion under climate change. *Nature Climate Change*, 6 (2), 166.

Hunt, B. G. (2011). Global characteristics of pluvial and dry multi-year episodes, with emphasis on megadroughts. *International Journal of Climatology*, 31 (10), 1425–1439.

Hunt, B. G., & Elliott, T. I. (2005). A simulation of the climatic conditions associated with the collapse of the Maya civilization. *Climatic Change*, 69 (2–3), 393–407.

Hurrell, J. W., Kushnir, Y., & Visbeck, M. (2001). The North Atlantic Oscillation. *Science*, 291 (5504), 603–605.

Hwang, C., Yang, Y., Kao, R., Han, J., Shum, C. K., Galloway, D. L., . . . & Li, F. (2016). Time-varying land subsidence detected by radar altimetry: California, Taiwan and north China. *Nature Scientific Reports*, 6, 28160.

Jackson, R. B., Carpenter, S. R., Dahm, C. N., McKnight, D. M., Naiman, R. J., Postel, S. L., & Running, S. W. (2001). Water in a changing world. *Ecological Applications*, 11 (4), 1027–1045.

Jasechko, S., Sharp, Z. D., Gibson, J. J., Birks, S. J., Yi, Y., & Fawcett, P. J. (2013). Terrestrial water fluxes dominated by transpiration. *Nature*, 496 (7445), 347.

Jean, B. (1985). Sécheresse et désertification au Mali. 2e partie: perspectives. *Revue Forestiere Française*, no. 4, 315–331.

Jouve, P. (1991). Sécheresse au Sahel et stratégies paysannes. *Sécheresse*, 2 (1), 61–69.

Juul, K. (1996). *Post drought migration and technological innovations among Fulani herders in Senegal: The triumph of the tube!* London: International Institute for Environment and Development.

Jyrkama, M. I., & Sykes, J. F. (2007). The impact of climate change on spatially varying groundwater recharge in the Grand River watershed (Ontario). *Journal of Hydrology*, 338 (3–4), 237–250.

Kaplan, J. O., Krumhardt, K. M., Ellis, E. C., Ruddiman, W. F., Lemmen, C., & Goldewijk, K. K. (2011). Holocene carbon emissions as a result of anthropogenic land cover change. *The Holocene*, 21 (5), 775–791.

Kasei, R., Diekkrüger, B., & Leemhuis, C. (2010). Drought frequency in the Volta Basin of West Africa. *Sustainability Science*, 5 (1), 89.

Kelley, C. P., Mohtadi, S., Cane, M. A., Seager, R., & Kushnir, Y. (2015). Climate change in the Fertile Crescent and implications of the recent Syrian drought. *Proceedings of the National Academy of Sciences*, 112 (11), 3241–3246.

Kintisch, E. (2016). Why did Greenland's Vikings disappear? *Science*. Retrieved from https://doi.org/10.1126/science.aal0363.

Kløve, B., Ala-Aho, P., Bertrand, G., Gurdak, J. J., Kupfersberger, H., Kværner, J., . . . & Uvo, C. B. (2014). Climate change impacts on groundwater and dependent ecosystems. *Journal of Hydrology*, 518, 250–266.

Knutti, R., & Sedláček, J. (2013). Robustness and uncertainties in the new CMIP5 climate model projections. *Nature Climate Change*, 3 (4), 369.

Konikow, L. F. (2011). Contribution of global groundwater depletion since 1900 to sea-level rise. *Geophysical Research Letters*, 38 (17).

Koplitz, S. N., Mickley, L. J., Marlier, M. E., Buonocore, J. J., Kim, P. S., Liu, T., . . . & Myers, S. S. (2016). Public health impacts of the severe haze in Equatorial Asia in September–October 2015: Demonstration of a new framework for informing fire management strategies to reduce downwind smoke exposure. *Environmental Research Letters*, 11 (9), 094023.

Koster, R. D., Dirmeyer, P. A., Guo, Z., Bonan, G., Chan, E., Cox, P., ... & Liu, P. (2004). Regions of strong coupling between soil moisture and precipitation. *Science*, 305 (5687), 1138–1140.

Krasting, J. P., Broccoli, A. J., Dixon, K. W., & Lanzante, J. R. (2013). Future changes in Northern Hemisphere snowfall. *Journal of Climate*, 26 (20), 7813–7828.

Kueppers, L. M., Snyder, M. A., & Sloan, L. C. (2007). Irrigation cooling effect: Regional climate forcing by land use change. *Geophysical Research Letters*, 34 (3).

Kuper, R., & Kröpelin, S. (2006). Climate-controlled Holocene occupation in the Sahara: Motor of Africa's evolution. *Science*, 313 (5788), 803–807.

Lamb, P. J. (1978a). Case studies of tropical Atlantic surface circulation patterns during recent sub-Saharan weather anomalies: 1967 and 1968. *Monthly Weather Review*, 106, 482–491.

Lamb, P. J. (1978b). Large-scale tropical Atlantic surface circulation patterns associated with sub-Saharan weather anomalies. *Tellus*, 30, 240–251.

Lamy, F., Hebbeln, D., Röhl, U., & Wefer, G. (2001). Holocene rainfall variability in southern Chile: A marine record of latitudinal shifts of the Southern Westerlies. *Earth and Planetary Science Letters*, 185 (3–4), 369–382.

Larrasoaña, J. C., Roberts, A. P., & Rohling, E. J. (2013). Dynamics of green Sahara periods and their role in hominin evolution. *PLoS One*, 8 (10), e76514.

Leathers, D. J., Yarnal, B., & Palecki, M. A. (1991). The Pacific/North American teleconnection pattern and United States climate: Part I, Regional temperature and precipitation associations. *Journal of Climate*, 4 (5), 517–528.

Leblanc, M. J., Favreau, G., Massuel, S., Tweed, S. O., Loireau, M., & Cappelaere, B. (2008). Land clearance and hydrological change in the Sahel: SW Niger. *Global and Planetary Change*, 61 (3–4), 135–150.

Lee, J. A., & Gill, T. E. (2015). Multiple causes of wind erosion in the Dust Bowl. *Aeolian Research*, 19, 15–36.

Levis, S., Bonan, G. B., & Bonfils, C. (2004). Soil feedback drives the mid-Holocene North African monsoon northward in fully coupled CCSM2 simulations with a dynamic vegetation model. *Climate Dynamics*, 23 (7–8), 791–802.

Li, Y., Wang, N. A., Zhou, X., Zhang, C., & Wang, Y. (2014). Synchronous or asynchronous Holocene Indian and East Asian summer monsoon evolution: A synthesis on Holocene Asian summer monsoon simulations, records and modern monsoon indices. *Global and Planetary Change*, 116, 30–40.

Liu, Z., Zhu, J., Rosenthal, Y., Zhang, X., Otto-Bliesner, B. L., Timmermann, A., . . . & Timm, O. E. (2014). The Holocene temperature conundrum. *Proceedings of the National Academy of Sciences*, 111 (34), E3501–E3505.

Lo, M. H., & Famiglietti, J. S. (2013). Irrigation in California's Central Valley strengthens the southwestern US water cycle. *Geophysical Research Letters*, 40 (2), 301–306.

Lobell, D., Bala, G., Mirin, A., Phillips, T., Maxwell, R., & Rotman, D. (2009). Regional differences in the influence of irrigation on climate. *Journal of Climate*, 22 (8), 2248–2255.

Lobell, D. B., & Bonfils, C. (2008). The effect of irrigation on regional temperatures: A spatial and temporal analysis of trends in California, 1934–2002. *Journal of Climate*, 21 (10), 2063–2071.

Lobell, D. B., Bonfils, C. J., Kueppers, L. M., & Snyder, M. A. (2008). Irrigation cooling effect on temperature and heat index extremes. *Geophysical Research Letters*, 35 (9).

Lu, X., Wang, L., & McCabe, M. F. (2016). Elevated CO_2 as a driver of global dryland greening. *Nature Scientific Reports*, 6, 20716.

Lurie, J. (2015, July 21). California's drought is so bad that thousands are living without running water. *Mother Jones*.

Luterbacher, J., & Pfister, C. (2015). The year without a summer. *Nature Geoscience*, 8 (4), 246.

Mahmood, R., Foster, S. A., Keeling, T., Hubbard, K. G., Carlson, C., & Leeper, R. (2006). Impacts of irrigation on 20th century temperature in the northern Great Plains. *Global and Planetary Change*, 54 (1), 1–18.

Mäkinen, R., Orvomaa, M., Veijalainen, N., & Huttunen, I. (2008). The climate change and groundwater regimes in Finland. In *Proceedings of the 11th International Specialized Conference on Watershed & River Basin Management*. Budapest: Trivent Publishing.

Malevich, S. B., & Woodhouse, C. A. (2017). Pacific sea surface temperatures, midlatitude atmospheric circulation, and widespread interannual anomalies in western US streamflow. *Geophysical Research Letters*, 44 (10), 5123–5132.

Maliva, R. G. (2014). Groundwater banking: Opportunities and management challenges. *Water Policy*, 16 (1), 144–156.

Manning, K., & Timpson, A. (2014). The demographic response to Holocene climate change in the Sahara. *Quaternary Science Reviews*, 101, 28–35.

Mantua, N. J., & Hare, S. R. (2002). The Pacific Decadal Oscillation. *Journal of Oceanography*, 58 (1), 35–44.

Marcott, S. A., Shakun, J. D., Clark, P. U., & Mix, A. C. (2013). A reconstruction of regional and global temperature for the past 11,300 years. *Science, 339* (6124), 1198–1201.

Margat, J., & van der Gun, J. (2013). *Groundwater around the world: A geographic synopsis.* London: CRC Press.

Margulis, S. A., Cortés, G., Girotto, M., & Durand, M. (2016). A Landsat-era Sierra Nevada snow reanalysis (1985–2015). *Journal of Hydrometeorology, 17* (4), 1203–1221.

Masson-Delmotte, V., Schulz, M., Abe-Ouchi, A., Beer, J., Ganopolski, A., González Rouco, J. F., . . . & Osborn, T. (2013). Information from paleoclimate archives. In T. F. Stocker, D. Qin, G.-K. Plattner, M. Tignor, S. K. Allen, J. Boschung, . . . & P. M. Midgley (Eds.), *Climate change 2013: The physical science basis: Working Group I contribution to the Fifth Assessment Report of the Intergovernmental Panel on Climate Change.* Cambridge, UK: Cambridge University Press.

Mattice, W. A. (1935). Dust storms, November 1933 to May 1934. *Monthly Weather Review, 63* (2), 53–55.

Maxwell, R. M., & Kollet, S. J. (2008). Interdependence of groundwater dynamics and land-energy feedbacks under climate change. *Nature Geoscience, 1* (10), 665.

McCabe, G. J., Palecki, M. A., & Betancourt, J. L. (2004). Pacific and Atlantic Ocean influences on multidecadal drought frequency in the United States. *Proceedings of the National Academy of Sciences, 101* (12), 4136–4141.

McCabe, G. J., & Wolock, D. M. (2016). Variability and trends in runoff efficiency in the conterminous United States. *JAWRA: Journal of the American Water Resources Association, 52* (5), 1046–1055.

McPhate, M. (2017, May 2). California today: Use less water, pay higher bills. *New York Times.*

Medellín-Azuara, J., MacEwan, D., Howitt, R. E., Sumner, D. A., Lund, J. R., Scheer, J., . . . & Kwon, A. (2016). *Economic analysis of the 2016 California drought on agriculture.* Davis: University of California–Davis, Center for Watershed Sciences.

Medina-Elizalde, M., & Rohling, E. J. (2012). Collapse of Classic Maya civilization related to modest reduction in precipitation. *Science, 335* (6071), 956–959.

Meko, D. M., Woodhouse, C. A., Baisan, C. A., Knight, T., Lukas, J. J., Hughes, M. K., & Salzer, M. W. (2007). Medieval drought in the upper Colorado River Basin. *Geophysical Research Letters, 34* (10).

Metcalfe, S. E., Barron, J. A., & Davies, S. J. (2015). The Holocene history of the North American monsoon: "Known knowns" and "known unknowns" in understanding its spatial and temporal complexity. *Quaternary Science Reviews, 120,* 1–27.

Miao, X., Mason, J. A., Swinehart, J. B., Loope, D. B., Hanson, P. R., Goble, R. J., & Liu, X. (2007). A 10,000 year record of dune activity, dust storms, and severe drought in the central Great Plains. *Geology, 35* (2), 119–122.

Middleton, N. J., & Thomas, D. S. (Eds.). (1997). *World atlas of desertification* (2nd ed.). London: Arnold.

Miller, G. H., Geirsdóttir, Á., Zhong, Y., Larsen, D. J., Otto Bliesner, B. L., Holland, M. M., . . . & Anderson, C. (2012). Abrupt onset of the Little Ice Age triggered by volcanism and sustained by sea-ice/ocean feedbacks. *Geophysical Research Letters, 39* (2).

Mote, P. W., Rupp, D. E., Li, S., Sharp, D. J., Otto, F., Uhe, P. F., . . . & Allen, M. R. (2016). Perspectives on the causes of exceptionally low 2015 snowpack in the western United States. *Geophysical Research Letters*, 43 (20).

Mote, P. W., Li, S., Lettenmaier, D. P., Xiao, M., & Engel, R. (2018). Dramatic declines in snowpack in the western US. *NPJ Climate and Atmospheric Science*, 1 (1), 2.

Mueller, B., & Seneviratne, S. I. (2012). Hot days induced by precipitation deficits at the global scale. *Proceedings of the National Academy of Sciences*, 109 (31), 12398–12403.

Munia, H., Guillaume, J. H. A., Mirumachi, N., Porkka, M., Wada, Y., & Kummu, M. (2016). Water stress in global transboundary river basins: Significance of upstream water use on downstream stress. *Environmental Research Letters*, 11 (1), 014002.

Musselman, K. N., Clark, M. P., Liu, C., Ikeda, K., & Rasmussen, R. (2017). Slower snowmelt in a warmer world. *Nature Climate Change*, 7 (3), 214.

Nash, D. J., De Cort, G., Chase, B. M., Verschuren, D., Nicholson, S. E., Shanahan, T. M., . . . & Grab, S. W. (2016). African hydroclimatic variability during the last 2000 years. *Quaternary Science Reviews*, 154, 1–22.

National Park Service. (2015). Storage capacity of Lake Mead. Retrieved from https://www.nps.gov/lake/learn/nature/storage-capacity-of-lake-mead.htm

Naumburg, E., Mata-Gonzalez, R., Hunter, R. G., Mclendon, T., & Martin, D. W. (2005). Phreatophytic vegetation and groundwater fluctuations: A review of current research and application of ecosystem response modeling with an emphasis on Great Basin vegetation. *Environmental Management*, 35 (6), 726–740.

Nicholson, S. E. (2013). The West African Sahel: A review of recent studies on the rainfall regime and its interannual variability. *ISRN Meteorology*, vol. 2013. Retrieved from https://doi.org/10.1155/2013/453521.

Nicholson, S. E., Dezfuli, A. K., & Klotter, D. (2012). A two-century precipitation dataset for the continent of Africa. *Bulletin of the American Meteorological Society*, 93 (8), 1219–1231.

Nicholson, S. E., Some, B., & Kone, B. (2000). An analysis of recent rainfall conditions in West Africa, including the rainy seasons of the 1997 El Niño and the 1998 La Niña years. *Journal of Climate*, 13 (14), 2628–2640.

Null, J. (2017). California's stressed water system: A primer. *Weatherwise*, 70 (1), 12–19.

Núñez, L., Grosjean, M., & Cartajena, I. (2002). Human occupations and climate change in the Puna de Atacama, Chile. *Science*, 298 (5594), 821–824.

Nyaga, J. M., Neff, J. C., & Cramer, M. D. (2015). The contribution of occult precipitation to nutrient deposition on the west coast of South Africa. *PloS One*, 10 (5), e0126225.

Obioha, E. E. (2009). Climate variability, environment change and food security nexus in Nigeria. *Journal of Human Ecology*, 26 (2), 107–121.

Oglesby, R. J., Sever, T. L., Saturno, W., Erickson, D. J., & Srikishen, J. (2010). Collapse of the Maya: Could deforestation have contributed? *Journal of Geophysical Research: Atmospheres*, 115 (D12).

O'Gorman, P. A. (2015). Precipitation extremes under climate change. *Current Climate Change Reports*, 1 (2), 49–59.

Oki, T., & Kanae, S. (2006). Global hydrological cycles and world water resources. *Science*, 313 (5790), 1068–1072.

Okkonen, J., & Kløve, B. (2010). A conceptual and statistical approach for the analysis of climate impact on ground water table fluctuation patterns in cold conditions. *Journal of Hydrology*, 388 (1–2), 1–12.

Okkonen, J., & Kløve, B. (2011). A sequential modelling approach to assess groundwater–surface water resources in a snow dominated region of Finland. *Journal of Hydrology*, 411 (1–2), 91–107.

Oldeman, L. R. (1992). Global extent of soil degradation. In *Bi-annual report 1991–1992, ISRIC* (pp. 19–36). Wageningen, Netherlands.

Oldeman, L. R., Hakkeling, R. U., & Sombroek, W. G. (1990). *World map of the status of human-induced soil degradation: An explanatory note*. Wageningen, Netherlands: International Soil Reference and Information Centre; Nairobi: United Nations Environment Programme.

Oldeman, L. R., & Van Lynden, G. W. J. (1997). Revisiting the GLASOD methodology. In R. Lal, W. E. H. Blum, C. Valentin, & B. A. Stewart (Eds.), *Methods for assessment of soil degradation* (pp. 423–439). Boca Raton, FL: CRC Press.

Olsson, L., Eklundh, L., & Ardö, J. (2005). A recent greening of the Sahel—Trends, patterns and potential causes. *Journal of Arid Environments*, 63 (3), 556–566.

Otkin, J. A., Anderson, M. C., Hain, C., Svoboda, M., Johnson, D., Mueller, R., . . . & Brown, J. (2016). Assessing the evolution of soil moisture and vegetation conditions during the 2012 United States flash drought. *Agricultural and Forest Meteorology*, 218, 230–242.

Otkin, J. A., Svoboda, M., Hunt, E. D., Ford, T. W., Anderson, M. C., Hain, C., & Basara, J. B. (2017). Flash droughts: A review and assessment of the challenges imposed by rapid onset droughts in the United States. *Bulletin of the American Meteorological Society*.

Palermo, E. (2015, July 23). An unexpected health consequence of the California drought. *LiveScience*.

Palmer, J. G., Cook, E. R., Turney, C. S., Allen, K., Fenwick, P., Cook, B. I., . . . & Baker, P. (2015). Drought variability in the eastern Australia and New Zealand summer drought atlas (ANZDA, CE 1500–2012) modulated by the Interdecadal Pacific Oscillation. *Environmental Research Letters*, 10 (12), 124002.

Pederson, N., Hessl, A. E., Baatarbileg, N., Anchukaitis, K. J., & Di Cosmo, N. (2014). Pluvials, droughts, the Mongol Empire, and modern Mongolia. *Proceedings of the National Academy of Sciences*, 111 (12), 4375–4379.

Polade, S. D., Pierce, D. W., Cayan, D. R., Gershunov, A., & Dettinger, M. D. (2014). The key role of dry days in changing regional climate and precipitation regimes. *Nature Scientific Reports*, 4, 4364.

Polgreen, L. (2007, February 11). In Niger, trees and crops turn back the desert. *New York Times*.

Postel, S. L., Daily, G. C., & Ehrlich, P. R. (1996). Human appropriation of renewable fresh water. *Science*, 271 (5250), 785–788.

Power, S., Casey, T., Folland, C., Colman, A., & Mehta, V. (1999). Inter-decadal modulation of the impact of ENSO on Australia. *Climate Dynamics*, 15 (5), 319–324.

Puma, M. J., & Cook, B. I. (2010). Effects of irrigation on global climate during the 20th century. *Journal of Geophysical Research: Atmospheres, 115* (D16).

Ramankutty, N., & Foley, J. A. (1999). Estimating historical changes in global land cover: Croplands from 1700 to 1992. *Global Biogeochemical Cycles, 13* (4), 997–1027.

Ravi, S., Breshears, D. D., Huxman, T. E., & D'Odorico, P. (2010). Land degradation in drylands: Interactions among hydrologic–aeolian erosion and vegetation dynamics. *Geomorphology, 116* (3–4), 236–245.

Reardon, T., Matlon, P., & Delgado, C. (1988). Coping with household-level food insecurity in drought-affected areas of Burkina Faso. *World Development, 16* (9), 1065–1074.

Richey, A. S., Thomas, B. F., Lo, M. H., Famiglietti, J. S., Swenson, S., & Rodell, M. (2015). Uncertainty in global groundwater storage estimates in a total groundwater stress framework. *Water Resources Research, 51* (7), 5198–5216.

Richey, A. S., Thomas, B. F., Lo, M. H., Reager, J. T., Famiglietti, J. S., Voss, K., . . . & Rodell, M. (2015). Quantifying renewable groundwater stress with GRACE. *Water Resources Research, 51* (7), 5217–5238.

Risbey, J. S., Pook, M. J., McIntosh, P. C., Wheeler, M. C., & Hendon, H. H. (2009). On the remote drivers of rainfall variability in Australia. *Monthly Weather Review, 137* (10), 3233–3253

Roberts, N., Brayshaw, D., Kuzucuoğlu, C., Perez, R., & Sadori, L. (2011). The mid-Holocene climatic transition in the Mediterranean: Causes and consequences. *The Holocene, 21* (1), 3–13.

Robock, A. (2000). Volcanic eruptions and climate. *Reviews of Geophysics, 38* (2), 191–219.

Rodgers, W. H. (2001). Executive orders and presidential commands: Presidents riding to the rescue of the environment. *Journal of Land Resources and Environmental Law, 21*, 13–24.

Rosen, W. (2014). *The third horseman: Climate change and the great famine of the 14th century.* New York: Penguin.

Roy, S. S., Mahmood, R., Niyogi, D., Lei, M., Foster, S. A., Hubbard, K. G., . . . & Pielke, R. (2007). Impacts of the agricultural Green Revolution–induced land use changes on air temperatures in India. *Journal of Geophysical Research: Atmospheres, 112* (D21).

Ruddiman, W. F. (2007). The early anthropogenic hypothesis: Challenges and responses. *Reviews of Geophysics, 45* (4).

Russell, H. C. (2009). *Climate of New South Wales: Descriptive, historical, and tabular.* Charleston, SC: BiblioBazaar. (Original work published 1877)

Sacks, W. J., Cook, B. I., Buenning, N., Levis, S., & Helkowski, J. H. (2009). Effects of global irrigation on the near-surface climate. *Climate Dynamics, 33* (2–3), 159–175.

Saini, R., Wang, G., & Pal, J. S. (2016). Role of soil moisture feedback in the development of extreme summer drought and flood in the United States. *Journal of Hydrometeorology, 17* (8), 2191–2207.

Saji, N. H., Goswami, B. N., Vinayachandran, P. N., & Yamagata, T. (1999). A dipole mode in the tropical Indian Ocean. *Nature, 401* (6751), 360.

Savtchenko, A. K., Huffman, G., & Vollmer, B. (2015). Assessment of precipitation anomalies in California using TRMM and MERRA data. *Journal of Geophysical Research: Atmospheres*, 120 (16), 8206–8215.

Scanlon, B. R., Faunt, C. C., Longuevergne, L., Reedy, R. C., Alley, W. M., McGuire, V. L., & McMahon, P. B. (2012). Groundwater depletion and sustainability of irrigation in the US High Plains and Central Valley. *Proceedings of the National Academy of Sciences*, 109 (24), 9320–9325.

Scanlon, B. R., Reedy, R. C., Stonestrom, D. A., Prudic, D. E., & Dennehy, K. F. (2005). Impact of land use and land cover change on groundwater recharge and quality in the southwestern US. *Global Change Biology*, 11 (10), 1577–1593.

Schlesinger, W. H., & Jasechko, S. (2014). Transpiration in the global water cycle. *Agricultural and Forest Meteorology*, 189, 115–117.

Schlesinger, W. H., Reynolds, J. F., Cunningham, G. L., Huenneke, L. F., Jarrell, W. M., Virginia, R. A., & Whitford, W. G. (1990). Biological feedbacks in global desertification. *Science*, 247 (4946), 1043–1048.

Schneider, U., Becker, A., Finger, P., Meyer-Christoffer, A., Rudolf, B., & Ziese, M. (2015). GPCC Full Data Reanalysis Version 7.0 at 0.5°: Monthly land-surface precipitation from rain-gauges built on GTS-based and historic data [Data set]. Retrieved from https://doi.org/10.5676/DWD_GPCC/FD_M_V7_050.

Schubert, S. D., Stewart, R. E., Wang, H., Barlow, M., Berbery, E. H., Cai, W., . . . & Mariotti, A. (2016). Global meteorological drought: A synthesis of current understanding with a focus on SST drivers of precipitation deficits. *Journal of Climate*, 29 (11), 3989–4019.

Schubert, S. D., Suarez, M. J., Pegion, P. J., Koster, R. D., & Bacmeister, J. T. (2004). On the cause of the 1930s Dust Bowl. *Science*, 303 (5665), 1855–1859.

Schubert, S. D., Wang, H., Koster, R. D., Suarez, M. J., & Groisman, P. Y. (2014). Northern Eurasian heat waves and droughts. *Journal of Climate*, 27 (9), 3169–3207.

Seager, R., Burgman, R., Kushnir, Y., Clement, A., Cook, E., Naik, N., & Miller, J. (2008). Tropical Pacific forcing of North American medieval megadroughts: Testing the concept with an atmosphere model forced by coral-reconstructed SSTs. *Journal of Climate*, 21 (23), 6175–6190.

Seager, R., Kushnir, Y., Ting, M., Cane, M., Naik, N., & Miller, J. (2008b). Would advance knowledge of 1930s SSTs have allowed prediction of the Dust Bowl drought?. *Journal of Climate*, 21 (13), 3261–3281.

Seager, R., Hoerling, M., Schubert, S., Wang, H., Lyon, B., Kumar, A., . . . & Henderson, N. (2015). Causes of the 2011–14 California drought. *Journal of Climate*, 28 (18), 6997–7024.

Seager, R., Ting, M., Davis, M., Cane, M., Naik, N., Nakamura, J., . . . & Stahle, D. W. (2009). Mexican drought: An observational modeling and tree ring study of variability and climate change. *Atmósfera*, 22 (1), 1–31.

Seager, R., Ting, M., Li, C., Naik, N., Cook, B., Nakamura, J., & Liu, H. (2013). Projections of declining surface-water availability for the southwestern United States. *Nature Climate Change*, 3 (5), 482.

Seneviratne, S. I., Corti, T., Davin, E. L., Hirschi, M., Jaeger, E. B., Lehner, I., . . . & Teuling, A. J. (2010). Investigating soil moisture–climate interactions in a changing climate: A review. *Earth-Science Reviews*, 99 (3–4), 125–161.

Shawki, D., Field, R. D., Tippett, M. K., Saharjo, B. H., Albar, I., Atmoko, D., & Voulgarakis, A. (2017). Long-lead prediction of the 2015 fire and haze episode in Indonesia. *Geophysical Research Letters*, 44 (19), 9996.

Sheets, H., & Morris, R. (1974). *Disaster in the desert: Failures of international relief in the West African drought*. Washington, DC: Carnegie Endowment for International Peace.

Shi, W., Tao, F., & Liu, J. (2014). Regional temperature change over the Huang-Huai-Hai Plain of China: The roles of irrigation versus urbanization. *International Journal of Climatology*, 34 (4), 1181–1195.

Shiklomanov, I. A. (2000). Appraisal and assessment of world water resources. *Water International*, 25 (1), 11–32.

Shukla, S. P., Puma, M. J., & Cook, B. I. (2014). The response of the South Asian summer monsoon circulation to intensified irrigation in global climate model simulations. *Climate Dynamics*, 42 (1–2), 21–36.

Shulmeister, J. (1999). Australasian evidence for mid-Holocene climate change implies precessional control of Walker circulation in the Pacific. *Quaternary International*, 57, 81–91.

Shuman, B. N., & Marsicek, J. (2016). The structure of Holocene climate change in mid-latitude North America. *Quaternary Science Reviews*, 141, 38–51

Siebert, S., Burke, J., Faures, J. M., Frenken, K., Hoogeveen, J., Döll, P., & Portmann, F. T. (2010). Groundwater use for irrigation—A global inventory. *Hydrology and Earth System Sciences*, 14 (10), 1863–1880.

Siebert, S., Döll, P., Hoogeveen, J., Faures, J.-M., Frenken, K., & Feick, S. (2005). Development and validation of the global map of irrigation areas. *Hydrology and Earth System Sciences*, 9 (5), 535–547.

Siebert, S., Kummu, M., Porkka, M., Döll, P., Ramankutty, N., & Scanlon, B. R. (2015). A global data set of the extent of irrigated land from 1900 to 2005. *Hydrology and Earth System Sciences*, 19 (3), 1521–1545.

Siegler, K. (2017, April 10). As California lifts drought restrictions, rural areas still lack running water. *All Things Considered* [Radio broadcast]. Washington, DC: National Public Radio.

Sivakumar, M. V. K. (2007). Interactions between climate and desertification. *Agricultural and Forest Meteorology*, 142 (2), 143–155.

Smirnov, O., Zhang, M., Xiao, T., Orbell, J., Lobben, A., & Gordon, J. (2016). The relative importance of climate change and population growth for exposure to future extreme droughts. *Climatic Change*, 138 (1–2), 41–53.

Smith, R. G., Knight, R., Chen, J., Reeves, J. A., Zebker, H. A., Farr, T., & Liu, Z. (2017). Estimating the permanent loss of groundwater storage in the southern San Joaquin Valley, California. *Water Resources Research*, 53 (3), 2133–2148.

Stahle, D. W., Cook, E. R., Burnette, D. J., Villanueva, J., Cerano, J., Burns, J. N., . . . & Szejner, P. (2016). The Mexican drought atlas: Tree-ring reconstructions of the soil moisture balance during the late pre-Hispanic, colonial, and modern eras. *Quaternary Science Reviews*, 149, 34–60.

Steponaitis, E., Andrews, A., McGee, D., Quade, J., Hsieh, Y. T., Broecker, W. S., . . . & Cheng, H. (2015). Mid-Holocene drying of the US Great Basin recorded in Nevada speleothems. *Quaternary Science Reviews*, 127, 174–185.

Stevens, M. (2016. November 18). 102 million dead California trees "unprecedented in our modern history," officials say. *Los Angeles Times*. Retrieved from http://www.latimes.com/local/lanow/la-me-dead-trees-20161118-story.html.

Stevenson, S., Timmermann, A., Chikamoto, Y., Langford, S., & DiNezio, P. (2015). Stochastically generated North American megadroughts. *Journal of Climate*, 28 (5), 1865–1880.

St. George, S., & Ault, T. R. (2014). The imprint of climate within Northern Hemisphere trees. *Quaternary Science Reviews*, 89, 1–4.

Stine, S. (1994). Extreme and persistent drought in California and Patagonia during mediaeval time. *Nature*, 369 (6481), 546.

Swain, D. L., Singh, D., Horton, D. E., Mankin, J. S., Ballard, T. C., & Diffenbaugh, N. S. (2017). Remote linkages to anomalous winter atmospheric ridging over the northeastern Pacific. *Journal of Geophysical Research: Atmospheres*, 122 (22).

Swain, D. L., Tsiang, M., Haugen, M., Singh, D., Charland, A., Rajaratnam, B., & Diffenbaugh, N. S. (2014). The extraordinary California drought of 2013/2014: Character, context, and the role of climate change. *Bulletin of the American Meteorological Society*, 95 (9), S3.

Swann, A. L., Fung, I. Y., Liu, Y., & Chiang, J. C. (2014). Remote vegetation feedbacks and the mid-Holocene Green Sahara. *Journal of Climate*, 27 (13), 4857–4870.

Swann, A. L., Hoffman, F. M., Koven, C. D., & Randerson, J. T. (2016). Plant responses to increasing CO_2 reduce estimates of climate impacts on drought severity. *Proceedings of the National Academy of Sciences*, 113 (36), 10019–10024.

Swinton, S. M. (1988). Drought survival tactics of subsistence farmers in Niger. *Human Ecology*, 16 (2), 123–144.

Tambo, J. A., & Abdoulaye, T. (2013). Smallholder farmers' perceptions of and adaptations to climate change in the Nigerian savanna. *Regional Environmental Change*, 13 (2), 375–388.

Taniguchi, M. (Ed.). (2011). *Groundwater and subsurface environments: Human impacts in Asian coastal cities*. Berlin: Springer.

Taylor, R. G., Scanlon, B., Döll, P., Rodell, M., Van Beek, R., Wada, Y., . . . & Konikow, L. (2013). Ground water and climate change. *Nature Climate Change*, 3 (4), 322.

Thomas, B. F., Famiglietti, J. S., Landerer, F. W., Wiese, D. N., Molotch, N. P., & Argus, D. F. (2017). GRACE groundwater drought index: Evaluation of California Central Valley groundwater drought. *Remote Sensing of Environment*, 198, 384–392.

Thompson, D. W., & Wallace, J. M. (2001). Regional climate impacts of the Northern Hemisphere annular mode. *Science*, 293 (5527), 85–89.

Tierney, J. E., Lewis, S. C., Cook, B. I., LeGrande, A. N., & Schmidt, G. A. (2011). Model, proxy and isotopic perspectives on the East African Humid Period. *Earth and Planetary Science Letters*, 307, 103–112.

Tierney, J. E., Pausata, F. S., & deMenocal, P. B. (2017). Rainfall regimes of the Green Sahara. *Science Advances*, 3 (1), e1601503.

Tierney, J. E., Smerdon, J. E., Anchukaitis, K. J., & Seager, R. (2013). Multidecadal variability in East African hydroclimate controlled by the Indian Ocean. *Nature*, 493 (7432), 389.

Tierney, J. E., & Ummenhofer, C. C. (2015). Past and future rainfall in the Horn of Africa. *Science Advances*, 1 (9), e1500682.

Trigo, R. M., Gouveia, C. M., & Barriopedro, D. (2010). The intense 2007–2009 drought in the Fertile Crescent: Impacts and associated atmospheric circulation. *Agricultural and Forest Meteorology*, 150 (9), 1245–1257.

Turral, H., Burke, J. J., & Faurès, J. M. (2011). *Climate change, water and food security* (Water Report No. 36). Rome: Food and Agriculture Organization of the United Nations.

UCAR (University Corporation for Atmospheric Research). (2011). Center for Science Education.

Van Beek, L. P. H., Wada, Y., & Bierkens, M. F. (2011). Global monthly water stress: 1. Water balance and water availability. *Water Resources Research*, 47 (7).

van der Molen, M. K., Dolman, A. J., Ciais, P., Eglin, T., Gobron, N., Law, B. E., . . . & Chen, T. (2011). Drought and ecosystem carbon cycling. *Agricultural and Forest Meteorology*, 151 (7), 765–773.

Van Loon, A. F. (2015). Hydrological drought explained. *Wiley Interdisciplinary Reviews: Water*, 2 (4), 359–392.

Van Loon, A. F., Gleeson, T., Clark, J., Van Dijk, A. I., Stahl, K., Hannaford, J., . . . & Hannah, D. M. (2016). Drought in the Anthropocene. *Nature Geoscience*, 9 (2), 89.

Van Loon, A. F., Tijdeman, E., Wanders, N., Van Lanen, H. A., Teuling, A. J., & Uijlenhoet, R. (2014). How climate seasonality modifies drought duration and deficit. *Journal of Geophysical Research: Atmospheres*, 119 (8), 4640–4656.

Veijalainen, N., Lotsari, E., Alho, P., Vehviläinen, B., & Käyhkö, J. (2010). National scale assessment of climate change impacts on flooding in Finland. *Journal of Hydrology*, 391 (3–4), 333–350.

Vrese, P., Hagemann, S., & Claussen, M. (2016). Asian irrigation, African rain: Remote impacts of irrigation. *Geophysical Research Letters*, 43 (8), 3737–3745.

Wada, Y., Beek, L. P., Sperna Weiland, F. C., Chao, B. F., Wu, Y. H., & Bierkens, M. F. (2012). Past and future contribution of global groundwater depletion to sea-level rise. *Geophysical Research Letters*, 39 (9).

Wada, Y., Wisser, D., & Bierkens, M. F. P. (2014). Global modeling of withdrawal, allocation and consumptive use of surface water and groundwater resources. *Earth System Dynamics*, 5 (1), 15.

Wada, Y., Wisser, D., Eisner, S., Flörke, M., Gerten, D., Haddeland, I., . . . & Tessler, Z. (2013). Multimodel projections and uncertainties of irrigation water demand under climate change. *Geophysical Research Letters*, 40 (17), 4626–4632.

Wang, C., Deser, C., Yu, J. Y., DiNezio, P., & Clement, A. (2017). El Nino and Southern Oscillation (ENSO): A review. In P. W. Glynn, D. P. Manzello, & I. C. Enochs (Eds.), *Coral reefs of the eastern tropical Pacific* (pp. 85–106). Dordrecht, Netherlands: Springer.

Wang, G., & Eltahir, E. A. (2000). Ecosystem dynamics and the Sahel drought. *Geophysical Research Letters*, 27 (6), 795–798.

Wang, J., Yang, B., Ljungqvist, F. C., Luterbacher, J., Osborn, T. J., Briffa, K. R., & Zorita, E. (2017). Internal and external forcing of multidecadal Atlantic climate variability over the past 1,200 years. *Nature Geoscience*, 10 (7), 512.

Wang, S. Y. S., Yoon, J. H., Gillies, R., & Hsu, H. H. (2017). The California drought: Trends and impacts. In S. Y. S. Wang, J. H. Yoon, C. C. Funk, & R. R. Gillies (Eds.), *Climate extremes: Patterns and mechanisms* (pp. 223–235). Hoboken, NJ: Wiley; Washington, DC: American Geophysical Union.

Wanner, H., Beer, J., Bütikofer, J., Crowley, T. J., Cubasch, U., Flückiger, J., . . . & Küttel, M. (2008). Mid- to late Holocene climate change: An overview. *Quaternary Science Reviews*, 27 (19–20), 1791–1828.

Wanner, H., Mercolli, L., Grosjean, M., & Ritz, S. P. (2015). Holocene climate variability and change: A data-based review. *Journal of the Geological Society*, 172 (2), 254–263.

Wanner, H., Solomina, O., Grosjean, M., Ritz, S. P., & Jetel, M. (2011). Structure and origin of Holocene cold events. *Quaternary Science Reviews*, 30 (21–22), 3109–3123.

Weathers, K. C. (1999). The importance of cloud and fog in the maintenance of ecosystems. *Trends in Ecology and Evolution*, 14 (6), 214–215.

Webb, P., & Reardon, T. (1992). Drought impact and household response in East and West Africa. *Quarterly Journal of International Agriculture*, 31, 230–24630.

Wei, Z., Yoshimura, K., Wang, L., Miralles, D. G., Jasechko, S., & Lee, X. (2017). Revisiting the contribution of transpiration to global terrestrial evapotranspiration. *Geophysical Research Letters*, 44 (6), 2792–2801.

Wezel, A., & Lykke, A. M. (2006). Woody vegetation change in Sahelian West Africa: Evidence from local knowledge. *Environment, Development and Sustainability*, 8 (4), 553–567.

Wilhite, D. A. (2000). Drought as a natural hazard: Concepts and definitions. In D. A. Wilhite (Ed.), *Drought: A global assessment* (Vol. I, pp. 3–18). London: Routledge.

Wilhite, D. A., & Glantz, M. H. (1985). Understanding the drought phenomenon: The role of definitions. *Water International*, 10 (3), 111–120.

Williams, A. P., Allen, C. D., Macalady, A. K., Griffin, D., Woodhouse, C. A., Meko, D. M., . . . & Dean, J. S. (2013). Temperature as a potent driver of regional forest drought stress and tree mortality. *Nature Climate Change*, 3 (3), 292–297.

Williams, A. P., Seager, R., Abatzoglou, J. T., Cook, B. I., Smerdon, J. E., & Cook, E. R. (2015). Contribution of anthropogenic warming to California drought during 2012–2014. *Geophysical Research Letters*, 42 (16), 6819–6828.

Wise, E. K. (2016). Five centuries of US West Coast drought: Occurrence, spatial distribution, and associated atmospheric circulation patterns. *Geophysical Research Letters*, 43 (9), 4539–4546.

Wisser, D., Fekete, B. M., Vörösmarty, C. J., & Schumann, A. H. (2010). Reconstructing 20th century global hydrography: A contribution to the Global Terrestrial Network-Hydrology (GTN-H). *Hydrology and Earth System Sciences*, 14 (1), 1.

Wolf, A. T., Yoffe, S. B., & Giordano, M. (2003). International waters: Identifying basins at risk. *Water Policy, 5* (1), 29–60.

Woodhouse, C. A., & Overpeck, J. T. (1998). 2000 years of drought variability in the central United States. *Bulletin of the American Meteorological Society, 79* (12), 2693–2714.

Worster, D. (2004). *Dust Bowl: The southern plains in the 1930s.* Oxford, UK: Oxford University Press.

Xiao, M., Koppa, A., Mekonnen, Z., Pagán, B. R., Zhan, S., Cao, Q., . . . & Lettenmaier, D. P. (2017). How much groundwater did California's Central Valley lose during the 2012–2016 drought? *Geophysical Research Letters, 44* (10), 4872–4879.

Xu, L., Shi, Z., Wang, Y., Chu, X., Yu, P., Xiong, W., . . . & Zhang, S. (2017). Agricultural irrigation-induced climatic effects: A case study in the middle and southern Loess Plateau area, China. *International Journal of Climatology, 37* (5), 2620–2632.

Xue, Y. (1996). The impact of desertification in the Mongolian and the Inner Mongolian grassland on the regional climate. *Journal of Climate, 9* (9), 2173–2189.

Yakirevich, A., Melloul, A., Sorek, S., Shaath, S., & Borisov, V. (1998). Simulation of seawater intrusion into the Khan Yunis area of the Gaza Strip coastal aquifer. *Hydrogeology Journal, 6* (4), 549–559.

Zeng, N., Neelin, J. D., Lau, K. M., & Tucker, C. J. (1999). Enhancement of interdecadal climate variability in the Sahel by vegetation interaction. *Science, 286* (5444), 1537–1540.

Zeng, N., & Yoon, J. (2009). Expansion of the world's deserts due to vegetation-albedo feedback under global warming. *Geophysical Research Letters, 36* (17).

Zhang, C. (2005). Madden-Julian Oscillation. *Reviews of Geophysics, 43* (2).

Zhao, M., & Running, S. W. (2010). Drought-induced reduction in global terrestrial net primary production from 2000 through 2009. *Science, 329* (5994), 940–943.

Ziese, M., Becker, A., Finger, P., Meyer-Christoffer, A., Rudolf, B., & Schneider, U. (2011). GPCC First Guess Product at 1.0°: Near real-time First Guess monthly land-surface precipitation from rain-gauges based on SYNOP data [Data set]. Retrieved from https://doi.org/10.5676/DWD_GPCC/FG_M_100.

Index

Printed in the USA
CPSIA information can be obtained
at www.ICGtesting.com
JSHW011518221024
72172JS00007B/55